PHOSPHOINOSITIDE 3-KINASE SIGNALLING PATHWAY

The Key to Cell Proliferation and Death

PHOSPHOINOSITIDE 3-KINASE SIGNALLING PATHWAY

The Key to Cell Proliferation and Death

Eric W-F Lam

Imperial College London, UK

ICP

Imperial College Press

Published by

Imperial College Press
57 Shelton Street
Covent Garden
London WC2H 9HE

Distributed by

World Scientific Publishing Co. Pte. Ltd.
5 Toh Tuck Link, Singapore 596224
USA office: 27 Warren Street, Suite 401-402, Hackensack, NJ 07601
UK office: 57 Shelton Street, Covent Garden, London WC2H 9HE

British Library Cataloguing-in-Publication Data
A catalogue record for this book is available from the British Library.

PHOSPHOINOSITIDE 3-KINASE SIGNALLING PATHWAY
The Key to Cell Proliferation and Death

ISBN 1-86094-626-7

Typeset by Stallion Press
Email: enquiries@stallionpress.com

Printed in Singapore by Mainland Press

To Anita and Oliver

Contents

Contributors

Eric W.-F. Lam
Department of Oncology
and Cancer Research-UK Labs
Division of SORA
MRC Cyclotron Building
Imperial College London
Hammersmith Campus
Du Cane Road
London W12 0NN, UK

Paul Brennan
Department of Medical Biochemistry
and Immunology
School of Medicine
Cardiff University
Henry Wellcome Research Building
Heath Park
Cardiff CF14 4XN, UK

Jan J. Brosens
Institute of Reproductive and
Developmental Biology
Division of SORA
MRC Cyclotron Building
Imperial College London
Hammersmith Campus
Du Cane Road
London W12 0NN, UK

Jan Domin
Division of Medicine
Imperial College London
Hammersmith Campus
Du Cane Road
London W12 0NN, UK

Silvia Fernandez de Mattos
Department of Oncology
and Cancer Research-UK Labs
Division of SORA
MRC Cyclotron Building
Imperial College London
Hammersmith Campus
Du Cane Road
London W12 0NN, UK

Andrew Sunters
Department of Oncology and Cancer
Research-UK Labs
Division of SORA
MRC Cyclotron Building
Imperial College London
Hammersmith Campus
Du Cane Road
London W12 0NN, UK

Foreword

This is an excellent book that provides a comprehensive and detailed review of phosphoinositide 3-kinases (PI3K) and their roles in controlling cell proliferation and apoptosis. Each chapter gives an up to date, authoritative overview of particular aspects of PI3K biology, that include structure, signal transduction, cell cycle control or the important downstream target of PI3K, the FOXO transcription factor. The book will be of interest to researchers who wish to understand particular aspects of PI3K biology in detail, as well as to students and academics who want a more broad understanding of this important and fast-changing field.

Shaun Thomas
King's College London
Rayne Institute, London
United Kingdom

Foreword

Over the last decade phosphatidylinositol 3-kinases (PI3Ks) have emerged as central players in the regulation of cellular function. Dysfunctional PI3K regulation has been reported in a plethora of diseases and their therapeutic relevance has recently become apparent. "*Phosphoinositide 3-kinase Siganlling Pathway*" is a well researched and timely publication that covers in detail our current knowledge concerning the regulation and function of this important family of lipid kinases. Each chapter approaches a different aspect of PI3K signaling, covering important topics including PKB (c-akt) function and the role of FOXO transcription factors in regulating cellular homeostasis. This book is an important reference work for all those who want an up-to-date review of this interesting signaling pathway.

Paul Coffer
University Medical Center Utrecht

1 Introduction

Eric W.-F. Lam

Tissue homeostasis in complex, multicellular organisms is dependent upon a dynamic equilibrium between cell proliferation, differentiation and death. These three vital processes ultimately determine cell number, tissue architecture and function. To preserve balance, cells are required to communicate constantly with one another, either directly or indirectly. Direct communications are mediated by cell-cell contact or by changes in the surrounding extracellular matrix. Indirect communications involve the secretion of growth factors, cytokines and hormones. To maintain tissue homeostasis, each cell within a tissue must decode the incoming signals and translate them into the correct cellular response.[1,2]

Binding of growth factor to its specific receptor, which is usually located on the cell surface, triggers a network of intracellular signals that eventually result in altered gene expression. This cascade of intracellular events upon activation of a receptor is called 'signal transduction'. Signal transduction is commonly mediated through a chain reaction that involves phosphorylation and dephosphorylation of relay proteins by kinases and phosphatases as well as by the release of second messengers such as inositol, calcium ions and diacylglycerol (DAG). Different signal transduction pathways interact with one another and it is this cross-talk that triggers the formation of larger, combinatorial signalling networks. For example, upon activation of their receptors, growth factors can activate a number of distinct, but intertwined, signalling cascades such as the JAK/STAT (Janus kinase/signal transducer and activator of transcription) the PI3K (phosphatidylinositol 3-kinase) and the MAPK (mitogen-activated protein kinase) pathways.[3–9]

Cells are therefore capable of responding differently to a stimulus, depending on which combinatorial signal transduction network is used to relay the signal to the nucleus.

The critical interplay between cell proliferation, death and differentiation is best exemplified by the haematopoietic system.[10,11] An adult human turns over 10^{11} blood cells daily by a process called programmed cell death. To compensate for this high rate of cell loss, the haematopoietic system must produce an equally high level of new blood cells every day. Moreover, increased production is required in times of physiological or pathological stress. For example, red blood cell numbers increase under conditions of hypoxia while granulocyte, macrophage and lymphocyte populations expand during infections. Perturbation of this equilibrium between cell production and elimination can lead to haematopoietic disorders, including autoimmune diseases, leukaemias and other myeloproliferative and lymphoproliferative disorders.

Cell proliferation requires activation of the cell cycle, a tightly regulated process that ensures that the cell is ready to divide.[11–13] The replicating DNA has to be free from damage to avoid mutation. Cellular events include duplication of DNA, condensation and segregation of the chromosomes and their symmetrical distribution into two daughter cells. The mammalian cell cycle can be divided into four phases: an S (DNA synthesis) phase and M (mitosis) separated by two intervals, G1 (gap 1 — between M and S) and G2 (gap 2 — between S and M) phases. The cell cycle is under stringent control during cell development and by restricting entry into the cell cycle stem cell and progenitor populations are protected from exhaustion. Conversely, unrestricted cell proliferation causes hyperplastic disorders and cancer.

Programmed cell death or apoptosis is a physiological process whereby superfluous cells commit suicide.[10,14] Apoptosis is an active, energy consuming process, characterised by a number of unique physiological features including loss of mitochondrial membrane potential, activation of the caspase protease cascades, membrane blebbing, condensation of the nucleus and the cytoplasm, and cleavage of proteins and DNA. *In vivo* these apoptotic cells will then be removed by macrophages or adjacent cells through a process termed autophagy. Depletion of growth factors and loss of appropriate cell-cell contact are common triggers for programmed cell death.

In addition, specific signals that activate death receptors, such as the Fas and TNF-receptors, can also induce apoptosis in target cells. A third way of inducing apoptosis is through cellular stress, including DNA damage, oxidative stress, viral infection and hypoxia.

The phosphoinositide 3-kinase (PI3K) signal transduction pathway has emerged as a critical regulator of cell fate decisions such as proliferation, differentiation and apoptosis.[3,5,6,8,9,14,15] Inhibition of this pathway invariably leads to cell cycle arrest and/or apoptosis. PI3Ks are a family of lipid kinases that serve as a nexus for signals generated by many different activated receptors and adhesion molecules. Once activated, PI3Ks generate phosphatidylinositols (PtdIns) (3,4,5) P_3 which lead to the recruitment and activation of PDK (3'-phosphoinositide-dependent kinase), Akt/PKB serine/threonine kinase, and G-proteins (e.g. Rac-GTPases). These intermediate molecules in turn regulate the activity of an ever-growing list of target proteins involved proliferation, survival and apoptosis. Downstream targets of the PI3K/Akt pathway include glycogen synthase kinase-3beta (GSK-3beta), mammalian target of rapamycin (mTOR), p70S6 kinase, endothelial nitric oxide synthase (eNOS) and several pro-apoptotic proteins, such as Bad, Bim and Caspase-9.

Perturbation in PI3K signalling underpins many diseases including heart disease, autoimmune/inflammatory disorders, cancer and resistance to chemotherapy.[3,6,8,9,15] Constitutive activation of the PI3K pathway is a consequence of enhanced expression of genes that encode class I PI3K subunits or Akt (PKB) genes or caused by genetic mutations in components of the pathway. For example, somatic deletions or mutations of *PTEN* (phosphatase and tensin homologue deleted on chromosome 10), an inhibitor of the PI3K/Akt pathway, have been identified in a large fraction (12%–60%) of human tumours of different histological origin. Consequently, considerable efforts are being made to develop new drugs that selectively target components of the PI3K signalling pathway for the treatment of cancers and other diseases.

This book provides a detailed overview of the recent advances in our understanding of PI3K signalling and its role in regulating cell proliferation, survival and apoptosis. Chapter 2 introduces the mechanisms and molecules that control cell proliferation and apoptosis. Chapter 3 provides an overview of the structural components of the PI3K signalling pathway while Chapter 4 discusses how this pathway functionally regulates cellular proliferation and apoptosis. The focus in Chapter 5 is on FOXO sub-family of forkhead transcription factors, the most recently uncovered downstream components of the PI3K signalling cascade. FOXO proteins are involved in the transcriptional regulation of many genes important for cell proliferation, cell death and differentiation.[3,6,8,9,15]

This book is intended for all those that are interested in cellular proliferation and apoptosis, whether a student or scientist. All authors responded enthusiastically when invited to contribute to this book and all have produced authoritative chapters which made editing an easy task.

References

1. H.S. Bell and K.M. Ryan, Intracellular signalling and cancer: complex pathways lead to multiple targets, *Eur. J. Cancer* **41** (2005) 206–215.
2. J.L. Bos and B.M. Burgering, Molecular mechanisms in signal transduction and cancer, *EMBO Rep.* **5** (2004) 855–859.
3. S. Brader and S.A. Eccles, Phosphoinositide 3-kinase signalling pathways in tumor progression, invasion and angiogenesis, *Tumori* **90** (2004) 2–8.
4. A. Chan, Ras-MAPK pathways, *Sci. STKE* **2005** (2005) tr5.
5. E. Foulstone, S. Prince, O. Zaccheo, J.L. Burns, J. Harper, C. Jacobs, D. Church and A.B. Hassan, Insulin-like growth factor ligands, receptors, and binding proteins in cancer, *J. Pathol.* **205** (2005) 145–153.
6. K. Okkenhaug and B. Vanhaesebroeck, PI3K in lymphocyte development, differentiation and activation, *Nat. Rev. Immunol.* **3** (2003) 317–330.
7. E.P. Reddy, A. Korapati, P. Chaturvedi and S. Rane, IL-3 signaling and the role of Src kinases, JAKs and STATs: a covert liaison unveiled, *Oncogene* **19** (2000) 2532–2547.
8. G. Song, G. Ouyang and S. Bao, The activation of Akt/PKB signaling pathway and cell survival, *J. Cell. Mol. Med.* **9** (2005) 59–71.
9. M.P. Wymann and R. Marone, Phosphoinositide 3-kinase in disease: timing, location, and scaffolding, *Curr. Opin. Cell Biol.* **17** (2005) 141–149.
10. G.J. Cowling and T.M. Dexter, Apoptosis in the haemopoietic system, *Philos. Trans. R. Soc. Lond. B Biol. Sci.* **345** (1994) 257–263.
11. R.A. Steinman, Cell cycle regulators and hematopoiesis, *Oncogene* **21** (2002) 3403–3413.
12. C.J. Sherr and J.M. Roberts, Living with or without cyclins and cyclin-dependent kinases. *Genes Dev.* **18** (2004) 2699–2711.

13. B.A. Weaver and D.W. Cleveland, Decoding the links between mitosis, cancer, and chemotherapy: the mitotic checkpoint, adaptation, and cell death. *Cancer Cell* **8** (2005) 7–12.
14. C.A. Lindemans and P.J. Coffer, Regulation of granulocyte apoptosis by phosphatidylinositol 3-kinase, *Biochem. Soc. Trans.* **32** (2004) 480–484.
15. P.J. Coffer, B.M. Burgering, Forkhead-box transcription factors and their role in the immune system, *Nat. Rev. Immunol.* **4** (2004) 889–899.

2 Cell proliferation and cell death

Andrew Sunters and Eric W.-F. Lam

1. Introduction

One of the basic challenges in biology is to identify the mechanisms whereby cell proliferation and cell death are regulated. The understanding of these processes will greatly enhance our knowledge of normal tissue homeostasis and may bring improvements in the treatment of diseases. At a very simplistic level, cell proliferation is required to build the biomass necessary for an organism to grow from an embryo to adulthood. Similarly, cell death is required to remove unwanted or damaged cells from tissues. Consequently, normal tissue homeostasis can be viewed as the maintenance of a fine balance between cell proliferation and cell death. This is particularly important in tissue and cell systems with a high rate of cell turnover such as during haematopoiesis. For example, an average healthy human adult turns over 10^{11} blood cells in a day. This represents a high volume of cell proliferation accompanied by an equally high level of cell death. Besides this routine turnover, haematopoietic cell proliferation is also required at times of physiological stress. For example, red blood cell numbers increase under conditions of hypoxia, while granulocyte, macrophage and lymphocyte populations expand during infections.

These cell proliferation and cell death mechanisms function in order respond to signals such as growth factors and hormones to shape solid tissue architecture of organs such as the liver, or regulate number of white cell such as macrophages or the lymphocyte compartment. Deregulation of these processes are often associated with disease states. The most dramatic examples of this is the deregulation of cell proliferation and lack of cell death seen in cancer and autoimmune lymphoproliferative syndrome (ALPS) and excessive cell death and lack of proliferation observed in degenerative disorders. This chapter will outline some of the key molecules and mechanisms involved in cell proliferation and cell death, and highlight their specialist function in specific tissues and diseases. Defects in the apoptotic signalling pathway are often associated with uncontrolled cell proliferation, high mutation rate and malignant transformation.

2. Cell Proliferation

In order for mammalian cells to multiply, they have to go through a process called the cell cycle. The mammalian cell cycle is divided into four distinct phases, historically these were defined as G_1 (**G**ap phase 1; where a cell prepares to undergo DNA replication), S (DNA **S**ynthesis; where the cell undergoes DNA synthesis), G_2 (**G**ap phase 2; where the cell prepares for mitosis) and M (**M**itosis when the newly replicated DNA condenses into sister chromatids which segregate to opposite ends of the cell, which then splits into two new daughter cells each containing an identical number of chromosomes). However, the majority of the cells in an adult mammal are out of the cell cycle, in a resting or quiescent state termed G_0. The molecular mechanisms which govern the initiation of cell division are complex and function to ensure that events which occur during cell division proceed in the correct temporal order. These regulatory pathways guard against cells uncontrollably replicating their DNA or mis-segregating their chromosomes, resulting in cells which have increased ploidy, or the incorrect number of individual chromosomes. Some of the key molecules that regulate cell division are cyclins, cyclin dependent kinases (CDKs) and cyclin dependent kinase inhibitors (CKIs). These proteins together make up the cell cycle machinery, which function in concert to ensure the cell cycle proceeds in the correct order.

2.1. Cyclins, CDKs and CKIs

The first cyclin to be identified was cyclin B, which was found to be part of the maturation promoting complex in sea urchins. The level of cyclin B increases as cells approached mitosis, but declines rapidly at the end of mitosis, and because of its 'oscillatory' expression it was named cyclin. Subsequent research has identified a large number of different cyclins, each with specific functions during the cell cycle. Cyclins associate with cyclin dependent kinases (CDKs) and form the regulatory subunit of CDK-cyclin holoenzyme complexes whose kinase activity is required for cell cycle progression.[1] The different cyclins have different affinities for distinct CDKs, for example D-type cyclins (cyclin D1, cyclin D2 and cyclin D3) bind preferentially to CDK4 and CDK6, cyclin E binds primarily to CDK2, cyclin A to CDK2 or CDK1 (also known as CDC2) and cyclin B to CDK1.[1] The different cyclins also function in different phases of the cell cycle, such that D-type cyclins and cyclin E function in G_1, cyclin A functions in S-phase and cyclin B functions in mitosis (Fig. 1). Further regulatory proteins called CKIs exist, which act to inhibit the CDK complexes. There are two families, the Cip/Kip family comprises $p21^{Waf1/Cip1}$, $p27^{Kip1}$ and $p57^{Kip2}$, and the INK4 family of p15INK4b, p16INK4a, p18INK4c and p19INK4d. The Cip/Kip family of CKIs have a broad range of substrate specificity, while the INK4 CKIs specifically target cyclin D-CDK4 or -6 activity. When over-expressed in cultured cells, all the INK4 family members inhibit cyclin D-CDK4/6 complexes and arrest cell cycle progression in G_1 in an pRb (retinoblastoma protein) dependent fashion. INK4 family members all contain at least four ankyrin repeats which bind tightly to CDK4 and CDK6 resulting in a much reduced binding affinity for both cyclin D and ATP, thereby reducing the catalytic activity of CDK4/6 in two ways. Interestingly, the INK4a locus also contains an alternative reading frame which expresses a protein called ARF — an 'alternative reading frame' encoding protein, which also regulates the cell cycle but by interacting with MDM2/HDM2 to prolong the half life of the p53 tumour suppressor.

The Cip/Kip family of CKIs are comprised of $p21^{Waf1/Cip1}$, $p27^{Kip1}$ and $p57^{Kip2}$. These proteins contain defined CDK and cyclin binding motifs in their amino terminus, and can bind to and inhibit not only cyclin D, but cyclin A and E-containing CDK complexes as well (Fig. 1). Initially it was thought that $p21^{Waf1/Cip1}$ and $p27^{Kip1}$ solely functioned to inhibit cyclin-CDK complexes, however it has become apparent that in the cytoplasm at low stoicheometries they can act as assembly factors for D-type cyclins

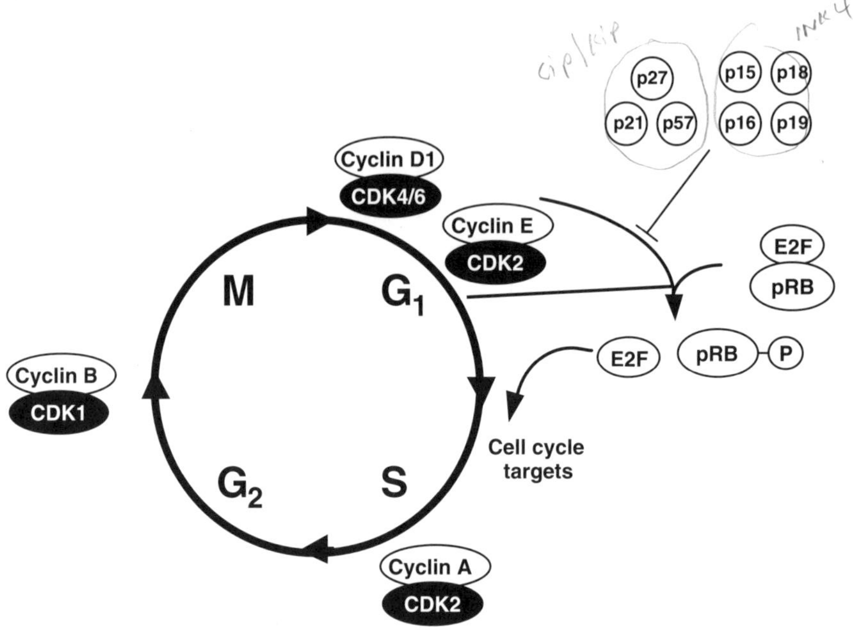

Fig. 1 Schematic representing the cell cycle. In G_1, D-type cyclins associate with CDK4 or 6 and cyclin E associates with CDK2 in order to phosphorylate the retinoblastoma protein (pRb). Phosphorylation of pRb results in the disassociation of pRb from the E2F transcription factor, allowing transcription of E2F target genes associated with S-phase induction such as those involved in DNA synthesis.[2,3] Cyclin A and CDK2 complexes are active during S-phase and contribute to the regulation of DNA synthesis. In G_2, cyclin B/CDC2 (CDK1) complexes function to promote mitosis. The cyclin-CDK complexes containing cyclins D, E, A and B can be inhibited by the CKIs from the Cip/Kip family (p21, p27 and p57), whilst the CD4/D-type cyclins can be inhibited by the members of the INK4 family (p15, p16, p18 and p19) in order to arrest cell cycle progression.

and CDK4/6 and their subsequent translocation into the nucleus.[2,3] At higher stoicheometries, $p21^{Waf1/Cip1}$ and $p27^{Kip1}$ act to inhibit D-type cyclin-CDK4/6 kinase activities. With respect to the complexes formed between cyclins A and E and CDK2, $p21^{Waf1/Cip1}$ and $p27^{Kip1}$ induce conformational changes in the CDKs which render the kinase inactive.[1]

2.2. The start of the cell cycle-G_0/G_1

Quiescent or resting cells in G_0 have no commitment to proliferation; however, stimulation of the cell by extracellular signals, such as growth factors, induces the cell to initiate a programme which results in cell proliferation.

During the period immediately following mitogenic signalling, the cell is not irreversibly committed to proliferation, but will monitor the availability of nutrients, and the strength and duration of mitogenic signals. As the cell passes from G_0 to G_1, they will reach a point in mid-G_1 termed the restriction point 'R'. Prior to 'R', withdrawal of mitogens will result in the cell failing to undergo cell division and returning to a quiescent state. After the restriction point, the cell is committed to cell division, and so mitogen withdrawal fails to arrest the cell cycle and the cell becomes resistant to growth inhibitory signals.

The early events of cell cycle progression are regulated by the D-type cyclins, cyclin D1, 2 and 3, and their cognate binding partners CDK4/6.[1] Many mitogenic signals result in the increased transcription and protein synthesis of D-type cyclins.[1,4–6] The increased levels of cyclin D result in increased levels of cyclin D-CDK4/6 complexes which become phosphorylated and activated by the CAK complex comprised of cyclin H-CDK9.

The active cyclin D-CDK4/6 complexes are then able to phosphorylate the key regulators of the restriction point in G_1; the retinoblastoma family of pocket proteins. The retinoblastoma (Rb) protein and its other family members, p107 and p130, function primarily to repress cell cycle progression[7] (Fig. 2). The mechanism by which Rb functions is by binding to transcription factors, such as the E2F family of proteins.[7] E2F activity is formed by heterodimers made up of an E2F protein (E2F1-5) and a DP protein (DP1-2). In its hypophosphorylated state, Rb associates with E2Fs and actively represses transcription of a subset of genes necessary for cell cycle progression.[8,9] This active repression is mediated by recruitment of sin3 and histone deacetylases (HDACs) which remodel the chromatin around the E2F target genes thereby blocking transcription[10–12] (Fig. 2). CyclinD-CDK4/6 complexes are capable of phosphorylating of Rb at the C-terminus,[13–15] and this phosphorylation leads to a displacement of HDAC activity from the Rb-E2F complex.[16] The effect of this is to remove the active repression of a subset of E2F target genes leaving only a degree of passive repression of other E2F targets caused by Rb.[16] One key E2F target gene whose expression is upregulated following the increase in cyclinD-CDK4/6 phosphorylation of Rb is *cyclin E*.[17,18] The expression of cyclin E becomes elevated during late G_1, and associates with its binding partner CDK2.[1]

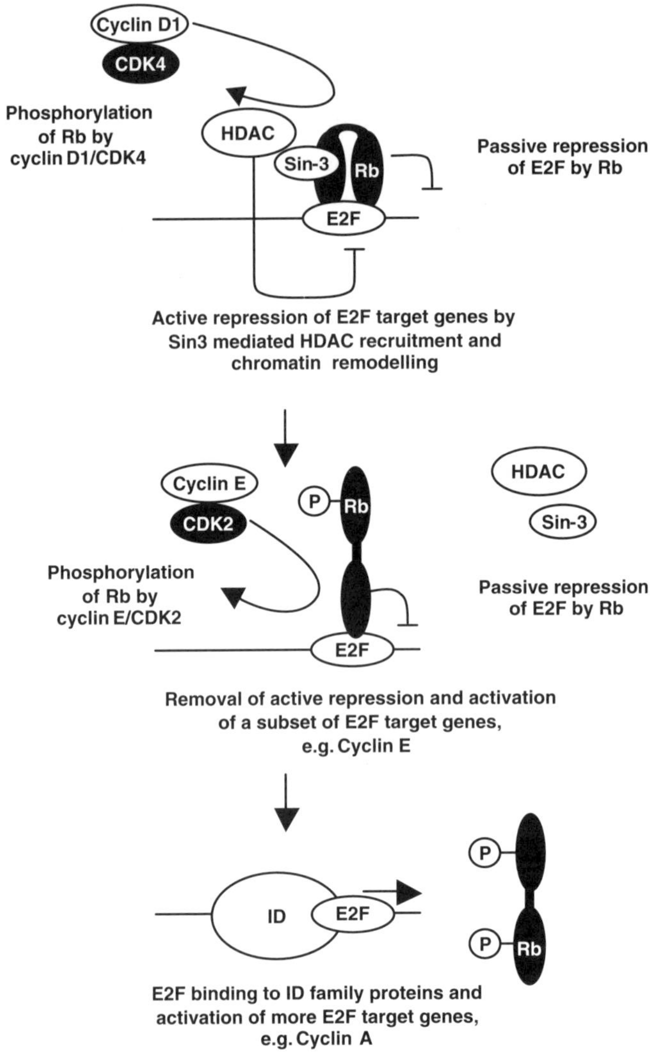

Fig. 2 The repression of E2F transcription is mediated by pRb. In its hypophosphorylated state pRb binds to E2F family members and is associated with the Sin3 protein. Sin3 is able to recruit histone de-acetylases (HDAC) which modify the local chromatin environment via de-acetylation in order to repress transcription. Phosphorylation of pRb by complexes containing D-type cyclins associated with either CDK4 or CDK6 results in the dissociation of Sin3 from pRb. The loss of HDAC activity results in the expression of some E2F target genes, although the binding of pRb to E2Fs is still able to repress E2F transcription of some target genes for example cyclin E, and the progression of the cells into late G_1. Further phosphorylation by cyclinE/CDK2 complexes results in dissociation of pRb from E2F, which allows E2F to bind to transcription activators of the ID family. This allows expression of target genes associated with S-phase such as cyclin A and enzymes involved in DNA biosynthesis.

Cyclin E-CDK2 also phosphorylate Rb family members.[1] Indeed it has been shown that the phosphorylation of Rb is sequential, with the initial phosphorylations being performed by D-type cyclin-CDK4/6 complexes which then induce conformational changes in Rb which results in enhanced phosphorylation by cyclin E-CDK2.[16] This further phosphorylation of Rb by cyclin E-CDK2 results in the dissociation of Rb from E2F family members.[1,16,19,20] E2F family members are then free to activate the transcription of a further set of genes required for cell cycle progression and DNA synthesis. One of these key E2F target genes that is required for cell cycle progression is cyclin A.[21] Cyclin A associates with either CDK2 or CDK1 in order to phosphorylate key targets required for successful S-phase traverse.

However, in order to prevent unscheduled entry into the cell cycle, the CKIs function to regulate CDK activity. In early G_0, the INK4 family members associate with CDK4 and 6 to prevent their association with D-type cyclins, whilst $p21^{Waf1/Cip1}$ and $p27^{Kip1}$ bind to the existing cyclin E and A complexes formed with CDK2 and CDK2 or 1, respectively. Higher order complexes of $p21^{Waf1/Cip1}$ and $p27^{Kip1}$ also bind to D-type cyclin complexes formed with CDK4/6 to inhibit CDK activity. As the cell is stimulated by mitogen, the levels of D-type cyclins increases, as well as some assembly factors such as SEI-1 which accelerates the formation of D-type cyclins and CDK4/6. As the activity of D-type cyclin/CDK complexes rises, and the levels of cyclin E rise, the higher *de novo* synthesised levels of D-type cyclins and CDK4/6 sequester the available $p21^{Waf1/Cip1}$ and $p27^{Kip1}$ which not only increases the assembly of D-type cyclin-CDK4/6 complexes, but also sequesters them away from *de novo* synthesised cyclin E-CDK2 complexes. This allows the activation of cyclin E-CDK2, and consequent increase in cyclin A synthesis, and thus S-phase progression.

Thus, the initiation of the cell cycle can be regarded as a highly orchestrated process involving increased transcription of D-type cyclins, which escape inhibition of INK4 family members, and sequester Cip-Kip family members in order to form complexes with CDK4/6, which then phosphorylate Rb family members, thereby allowing the expression of cyclin E. Complexes formed between cyclin E and CDK2 further phosphorylate Rb family members, removing the inhibitory effects on E2F family members by the Rb family, and allowing the increased expression of cyclin A and other S-phase targets.

2.3. S-phase-DNA synthesis

The full activation of E2F family members allows for the expression of several classes of genes that are necessary for S-phase traverse. These include proteins that are required to ensure that the pools of nucleotides are sufficient for effective DNA replication, such as thymidine kinase, thymidilate kinase, dihydrofolate reductase and ribonucleotide reductase. Furthermore, E2F activation is needed for the expression of some of the components of the DNA replication machinery, including histone H2a, DNA polymerase α as well as replication protein A (RPA2), CDC6, Orc1 and MCM 2, 3, 4, 5, 6 and 7 (Fig. 2).

The cyclin-Rb-E2F pathway in G_1 functions in order to ensure that cells enter S-phase only when stimulated to do so, cyclin-CDKs also act to prevent more than one round of semi-conservative DNA replication. In G_1, the origin recognition complex (ORC, comprised of Orc 1–6 subunits) binds to replication origins in the chromatin.[22,23] Orc1 in association with newly synthesised CDC6 then binds to the DNA-bound ORC complex, and this is then capable of recruiting Ctd1 to the ORC complex (Fig. 3). Ctd1/CDC6 then recruit a complex comprised of newly synthesised MCM 2–7 to the ORC complex to form the pre-replication complex.[22] The function of MCM 2–7 is to act as a helicase in order to unwind the DNA prior to DNA synthesis. As S-phase commences, MCM10 binds to the pre-replication complex.[22] This allows cyclin A-CDK2 to phosphorylate CDC6, causing it to be released from the complex.[24] Cyclin E-CDK2 phosphorylates one or more proteins in the PRC, thereby allowing Cdc45, which in its phosphorylated form recruits the DNA polymerase complex to the origin to begin DNA replication, and S-phase begins in earnest. Cyclin A-CDK2 also phosphorylates DNA polymerases. Ctd1 is inactivated by a protein called Gemin which leads to its nuclear export and degradation.[25] The phosphorylation of Cdc6 by cyclin A-CDK2 also leads to its export from the nucleus and eventual degradation.[24] Consequently, the removal of Cdc6 from the PRC by cyclin A-CDK2 activity, and the fact that cyclin E is degraded shortly after S-phase entry, helps to prevent the formation of new PRCs and the recruitment of polymerases to ORCs.[23] This has the effect of preventing more than one round of DNA synthesis from any one origin, thereby ensuring that all the genetic material is replicated once and only once.[23] After the genome has been replicated, the cell then passes into the G_2 phase.

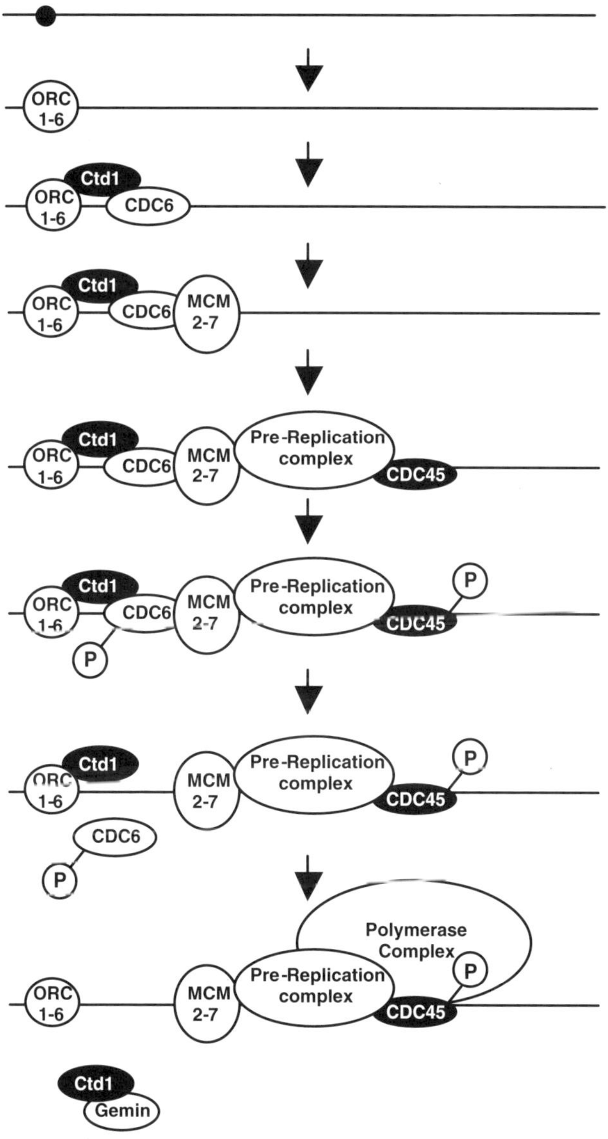

Fig. 3 Schematic representing the control of DNA replication during the S-phase. The origins of replication recruit the origin of replication complex (ORC) comprised of Orc 1–6 proteins. The ORC then recruits newly synthesised cdc6 and Ctd1. Ctd1/CDC6 then recruits newly synthesised MCM 2–7 to the ORC complex to form the pre-replication complex. MCM 2–7 act as a helicase unwinding the DNA prior to DNA polymerisation and recruits the pre-replication complex (PRC). As S-phase starts, cyclin A-CDK2 phosphorylates CDC6, causing its dissociation from the complex. Cyclin E-CDK2 phosphorylates proteins in the PRC, thereby recruiting Cdc45, which is then phosphorylated before recruiting the DNA polymerase complex to the origin to begin DNA synthesis.

2.4. G_2 and mitosis

One of the most essential proteins required for mitosis is cyclin B. During S-phase there is increased transcription of the *cyclin B* gene. Cyclin B protein associates with its cognate kinase-binding partner CDK1 in the cytoplasm.[1] In S and G_2 phase of the cell cycle, CDK1 is inactivated by phosphorylation on Thr 14 and Tyr15 residues by the kinases Myt-1[26] and Wee1,[27] respectively, but activated by CAK phosphorylation of Thr 161.[28] Cyclin B contains both a cytoplasmic retention and a nuclear localisation sequence, and phosphorylations in these regions regulate the localisation of the cyclin B-CDK1 complex.[29] (Fig. 4). Throughout G_2 cyclin B-CDK1 is located in the cytoplasm, but sequential phosphorylations by ERK2 and polo-like kinase (Plk1) within the CRS result in the translocation of cyclin B to the nucleus.[29] Furthermore, dephosphorylation of the Wee1 and Myt-1 phosphorylation sites by CDC25c result in the activation of CDK1 kinase.[29,30] CDC25 itself is also maintained in the cytosol until phosphorylated by PLK.[31,32] Consequently active cyclin B-CDK1 dimers move into the nucleus and induce

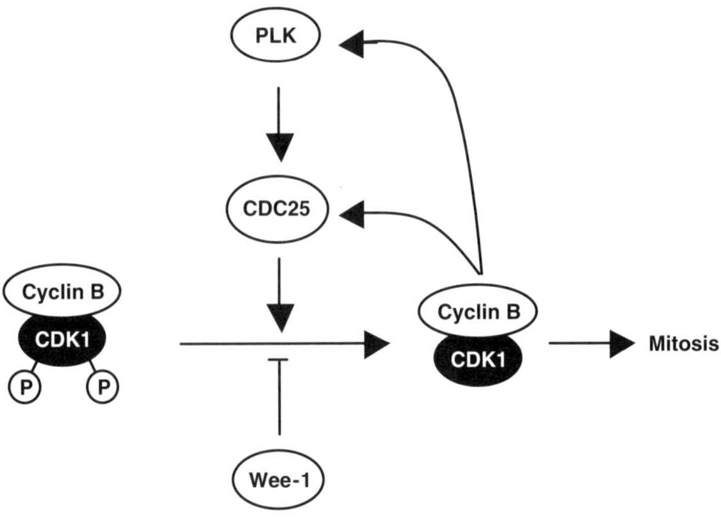

Fig. 4 The activation of cyclin B/cdc2 complexes is mediated by the phosphorylation of cdc2 at Y15 and Y17 by Wee-1 and Myt1. The action of polo-like kinases (PLKs) activates the cdc25c phosphatase which removes these inhibitory phosphorylations on cdc2 thereby activating it. Active cyclin B/cdc2 complexes also activate both PLKs and cdc25c in a feed-forward loop, enhancing their own activation leading to mitosis.

mitotic progression by phosphorylating key substrates involved in chromatin condensation and lamin regulation, which results in the breakdown of the nuclear membrane.[33,34] Cyclin B-CDK2 also regulates the microtubules which act to separate the sister chromatids, as well as the actin cytoskeleton which results in the cell becoming more rounded, thereby allowing the two daughter cells to separate effectively.

Another key player in mitosis is the centrosome. This is a multiprotein complex which organises the microtubules required to segregate the sister chromatids. In newly divided daughter cells there is only one centrosome, but this becomes duplicated prior to mitosis. During G_2 the newly duplicated pair of centrosomes are closely associated and found in the cytoplasm where they act to organise microtubule formation.[35] During prophase, the two centrosomes migrate to opposite ends of the cell, either side of the nucleus in order to coordinate the future spindle poles (Fig. 5). The centrosome subunit protein C-Nap1 which acts to keep the paired centrosomes together is phosphorylated, and inactivated by members of the NIMA family of kinases allowing separation of the two centrosomes.[36,37] As they separate, the centrosomes recruit additional multiprotein structures called G-tubulin ring complexes which act as additional nucleation sites for microtubule assembly.[35] Cyclin B-CDK1 phosphorylate key kinesin-related motor proteins which then bind to the centrosomes. These then interact with cytoplasmic dyneins in order to pull the centrosomes apart (Fig. 5).

Once the centrosomes have been separated, the chromatin begins to condense. The precise mechanism of DNA condensation remains unclear, but is known to involve modification of the nuclear proteins, including histones, by phosphorylation, methylation and acetylation. In order for the chromatin to fully condense, the extent of negative supercoiling is increased by topoisomerase II and a multiprotein complex called condensin.[35] There is again some evidence that cyclin B-CDK1 functions in the regulation of these proteins. Ultimately, the DNA condenses to form sister chromatids joined at the centromeres. Once the centromeres have separated, the nuclear membrane also begins to breakdown under the influence of cyclin B-CDK1 (Fig. 5).

At this point, prometaphase, the centrosomes extend microtubule filaments or spindles fibres, throughout the cell until they encounter kinetocores. Kinetocores are multiprotein complexes which are associated with the centromeres of the sister chromatids. These attachments are transient in

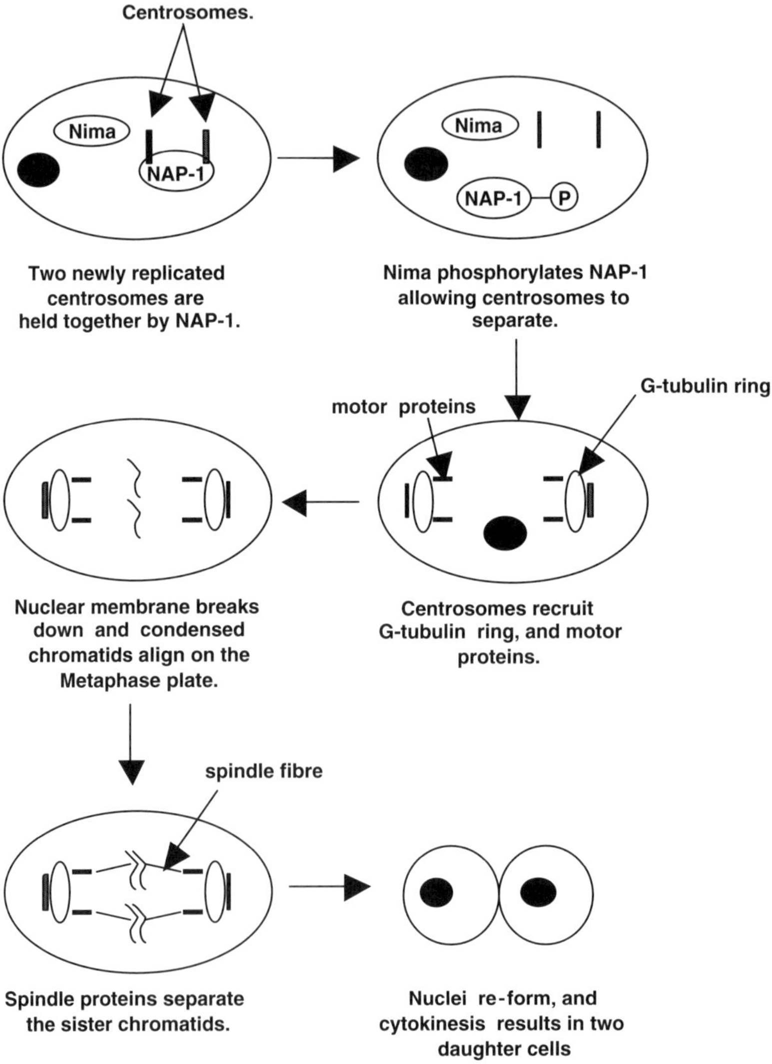

Fig. 5 Schematic illustrating the events leading to mitosis. As cells progress towards mitosis, the newly replicated centrosomes are held together by the NAP-1 protein. Phosphorylation of NAP-1 by NIMA kinases results in the separation of the centrosomes and their migration to opposite ends of the cell. Recruitment of G-actin rings and motor proteins by the centrosomes allows the formation of the spindle apparatus. Following the breakdown of the nuclear membrane, the replicated chromatids condense and line up on the metaphase plate where the spindle fibres bind to the centromeres. During anaphase, the spindle fibres pull the sister chromatids apart, and the nuclear membrane reforms. Two new daughter cells are produced following cytokinesis.

nature, and the spindles can attach, detach then reattach to the kinetocore. The motors in the spindle fibres from each spindle pole push and pull the sister chromatids. This jostling continues until there is a definite 'bi-polar' attachment, i.e. there are spindles originating from both spindle poles (centromeres) attached to one of the centromeres of each sister chromatids.[38] At this point, called metaphase, the chromatids are aligned in the middle of the cell on the metaphase plate.

Also during metaphase, a multiprotein complex comprised of at least 11 subunits with ubiquitin ligase activity called the anaphase promoting complex (APC) plays a key role. APC is only active as a ubiquitin ligase when it is associated with either Cdc20 or Cdh1. High levels of CDK1 activity, either associated with cyclin A or B allows the binding of Cdc20, which activates the ubiquitin ligase activity (Fig. 6). APCcdc20 then ubiquitinates cyclin A during late prometaphase and metaphase, as well as cyclin B during metaphase and anaphase.[38] Once the sister chromatids have condensed during prophase, they are held together at the centromere by a protein called cohesin.[39] Cohesin is a substrate for a serine protease called separase (Fig. 6). To prevent premature anaphase, separase is inactivated by binding to a protein called securin, and by phosphorylation by cyclin B-CDK1. During anaphase, the APCcdc20 complex ubiquitinates both securin and cyclin B, thereby allowing separase to firstly dissociate from securin, and secondly, in the absence of cyclin B (and therefore CDK1 activity), be subject to phosphatase activity.[38] This allows separase to become fully active, and degrade the cohesin which holds the sister chromatids together.[39] With the loss of cohesin, the sister chromatids are no longer held together at the centromeres, and are pulled apart by the spindles[38,39] (Fig. 6). This occurs by a combination of motor proteins and tubule shrinkage. It has been shown that in Xenopus XKid protein, which is a kinesin motor protein, that forces the chromatids away from the spindle poles during the chromatid oscillation during prometaphase. This is also a substrate for ubiquitination and consequent degradation for the Xenopus APC-Cdc20, thereby allowing sister chromatid separation.[38]

As the sister chromatids reach the poles of the cell during the next phase of mitosis, telophase, the nuclear membrane begins to reform, and the chromatids begin to decondense and the nucleus starts to reform. During this phase, a contractile ring composed of actinomysin fibres begins to

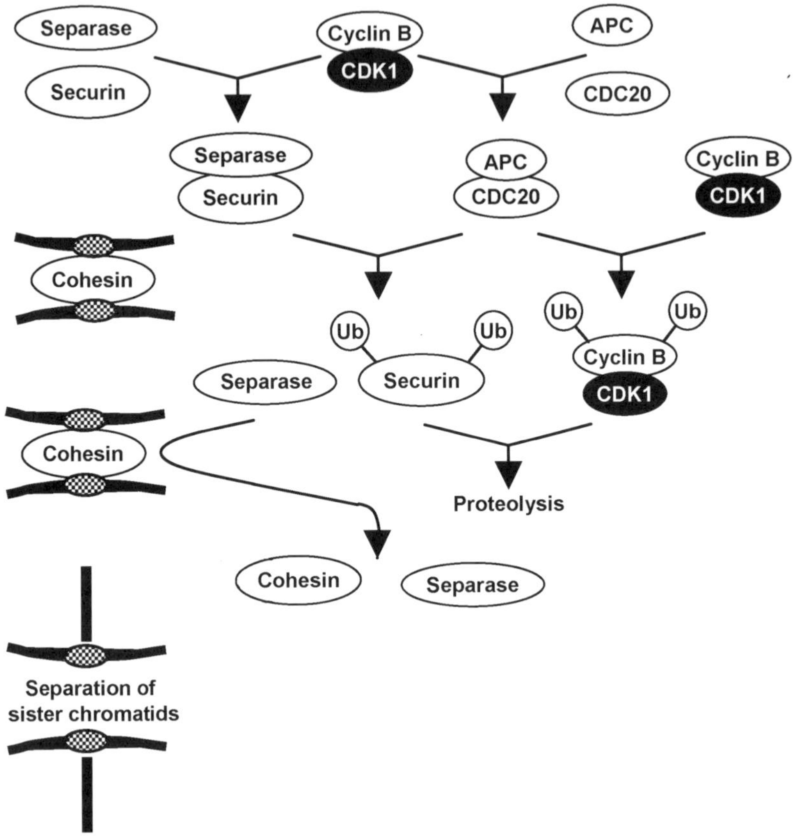

Fig. 6 Regulation of sister chromatid separation. The action of cyclin B/cdc2 complexes results in the formation of complexes between separase and securin as well as between APC and cdc20. The ubiquitination activity of APC-cdc20 complexes results in the ubiquitination and subsequent proteolysis of securin and cyclin B. The activated separase is the able to proteolytically cleave cohesin, a protein which holds the centromeres of the sister chromatids together. The loss of cohesin results in the separation of the sister chromatids during anaphase.

constrict the cell and cytokinesis occurs, resulting in the separation of two new daughter cells. During anaphase and telophase, the Cdc20 subunit of APC is replaced by the Cdh1 protein, which then ubiquitinates the Cdc20 subunit, targeting it for destruction.[38] The APCCdh1 complex then persists throughout the following G_0/G_1 phase of the cell cycle. APCCdh1 is able to degrade cyclin D1 (but not D2 or D3), A and B, but is inactive from the start of S phase, when the Cdh1 component is heavily phosphorylated by cyclin E-, A- and B-associated kinase activity.[38]

2.5. Cell cycle checkpoints

DNA replication is a very important and highly coordinated process, and misreplication of DNA can have catastrophic effects on the cell. Consequently, so-called DNA damage checkpoint pathways exist which are able to respond to DNA damage and cellular insults in order to arrest the cell cycle. The arrest of the cell cycle allows the various DNA repair pathways to repair the genome, thereby preventing the accumulation of mutations in daughter cells. There are several pathways that can transduce the signal from damaged DNA to the cell cycle, and they function at either the G_1/S, S and G_2/M phases of the cell cycle, as well as a post-mitotic or decatenation checkpoint.

2.6. The G_1/S checkpoint

DNA damage which occurs during G_0/G_1, such as DNA strand breaks caused by radiation, causes the ATM (ataxia telangectasia mutated) protein to become activated. Normally, ATM is present in the cell as a dimer, but DNA damage stimulates its kinase activity, resulting in autophosphorylation, and separation into active monomers.[40] ATM then migrates to sites of DNA damage and phosphorylates a number of substrates within the cell such as NBS1 (Nijmegen breakage syndrome 1), BRCA1 (breast cancer 1) and the tumour suppressor p53[41] (Fig. 7). Active ATM also phosphorylates and activates the key mediator of cell cycle arrest, CHK2.[42] This protein phosphorylates p53 in the transcriptional regulatory domain, increasing its activity, as well as the ubiquitin ligase MDM2.[43] Normally, MDM2 ubiquitinates p53, resulting in its degradation.[44] However, CHK2 and ATM phosphorylation inhibits the MDM2-dependent degradation of p53, thereby providing a mechanism whereby p53 protein is both stabilised and activated, resulting in increased expression of p53 target genes. One of the key targets of p53 is the CKI, $p21^{Waf1/Cip1}$.[45] The resultant increased levels of $p21^{Waf1/Cip1}$ inhibit cyclin E-CDK2 activity and thus phosphorylation of Rb family members and activation of E2F proteins, thereby halting cell cycle progression (Fig. 7). p53 also transcriptionally represses the

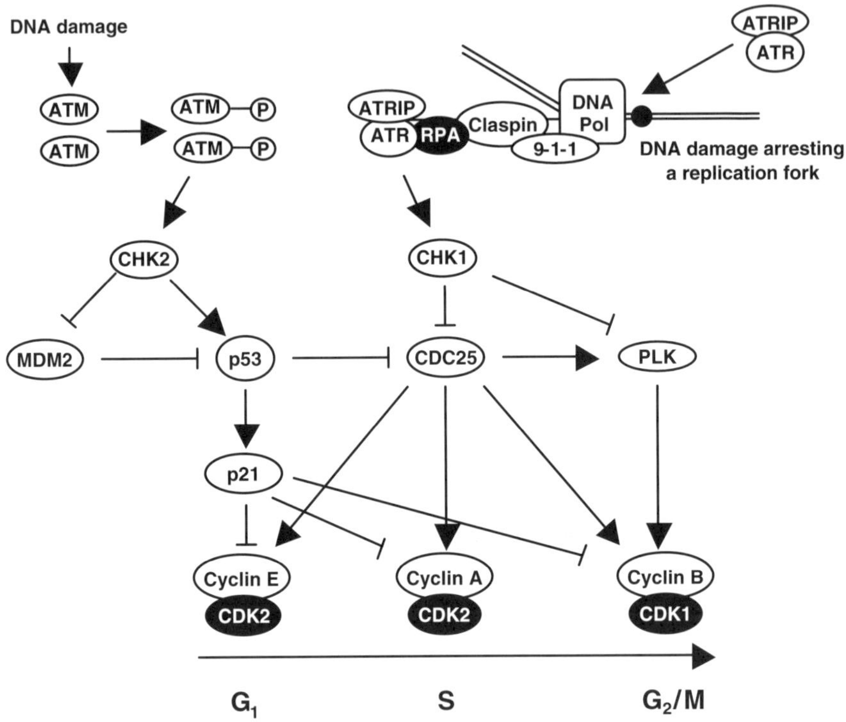

Fig. 7 Schematic representing the cell cycle checkpoints. DNA damage stimulates the auto-phosphorylation of ATM resulting in its activation. ATM then phosphorylates CHK2 which activates it, and ATM in turn phosphorylates MDM2 to inactivate it and inhibit its p53 ubiquitination ability, thereby stabilising p53. CHK2 also phosphorylates p53 directly, and this increases its transcriptional activity. p53 is then able to activate transcription of the p21 gene resulting in the inhibition of cyclin/CDK complexes, this is the G_1 checkpoint. In the S-phase checkpoint, p53-mediated p21 activation slows down the cell cycle via inhibition of cyclin A, but p53 is also capable of inhibiting cdc25 family members, thereby inhibiting CDK activation. Furthermore, replication forks stalled by DNA damage are able to recruit the ATR protein which activates the CHK1 kinase. CHK2 also inhibits cdc25 proteins preventing CDK2 activation and S-phase progression. In the G_2 checkpoint, cyclin B/cdc2 complexes are inhibited in a multifactoral manner to prevent the onset of mitosis with a genome containing DNA damage. Not only is p21 capable of inhibiting cyclin B/cdc2, but the inhibition of cdc25 by p53 and CHK2 prevents the activation of cdc2 activity. Furthermore, CHK2 also inactivates PLK family members by phosphorylation, thereby preventing mitotic progression.

expression of the cdc25c phosphatase which activates CDK2, providing an additional mechanism for the induction of G1 arrest.[46] These mechanisms function to ensure that the cell does not attempt to commence DNA replication in the presence of DNA damage, however additional pathways function to arrest the cell cycle once DNA synthesis has started.

2.7. S-phase checkpoint

Whilst the activation of the ATM-CHK-p53-p21$^{Waf1/Cip1}$ pathway causes cell cycle arrest in G_0/G_1, it also functions to arrest the cell cycle in S-phase. However there are also other mechanisms whereby DNA synthesis can be halted. One of the key features of S-phase is the presence of replication forks. DNA strand breaks, as well as some DNA damaging agents that covalently bind to DNA are able to block processive RNA and DNA polymerases. One protein that responds to stalled replication forks caused by this type of damage is ATR (Ataxia telangectasia and RAD3-related). ATR is normally bound to ATRIP (ATR interacting protcin).[47] Following stalling of the replication fork, ATR is recruited via the interaction between ATRIP and the single-stranded DNA-binding protein RPA (Fig. 7).[48] Further proteins need to be recruited to the site, such as RSR (RAD17 containing complex) and claspin which load the sliding clamp complex 9-1-1 (containing RAD9, HUS-1 and RAD1).[49] The assembly of this multiprotein complex serves not only to stimulate the activity of ATR, but also to sheath the single-stranded DNA located at the stalled replication fork, thereby protecting it from damage.[49] After this multiprotein complex is assembled, ATR is able to phosphorylate CHK1. CHK1, like CHK2, phosphorylates CDC25, resulting in its degradation. The loss of CDC25a results in a failure to activate CDK2 which is required for the efficient loading of CDC45 onto chromatin, and thus the recruitment of DNA polymerase onto new origins of replication.[42] Also during S-phase, ATM is able to phosphorylate SMC1, a protein which is a member of the cohesin family, and which helps to facilitate DNA repair.[41] Thus these pathways function in S-phase to inhibit cell cycle progression, prevent DNA synthesis, protect the vulnerable DNA at the replication fork and promote DNA repair.

2.8. The G_2/M checkpoint

After cells have undergone DNA replication, they undergo mitosis to produce two new daughter cells. However, if the genome possesses any unrepaired DNA lesions or breaks, then mitosis will result in an aberrant mitosis. Consequently, the G_2/M checkpoint functions to prevent mitosis in cells containing DNA damage. The primary target of G_2/M arrest is the cyclin B/CDK1 complex. Activation of p53 via ATM leads to increased expression of $p21^{Waf1/Cip1}$ which is able to inhibit the cyclin B-CDK1 complex[50] (Fig. 7). However, in cell lines grown *in vitro* it has been shown that whilst loss or mutation of the p53 tumour suppressor is sufficient to abrogate the $p21^{Waf1/Cip1}$-mediated G1 block, it does not affect the G_2/M arrest. This would suggest that p53 is not necessary for the G_2/M arrest and that other mechanisms function to induce the G_2/M arrest. Inhibition of cyclin B-CDK1 is mediated via inhibition of CDC25C by the ATM/ATR-CHK2/CHK1 pathway.[51] In addition, the polo-like kinase family members PLK1 and PLK3, which promote the onset of M phase, are also down-regulated by the ATM/ATR-CHK2/CHK1 pathway.[52]

2.9. Additional checkpoint controls

In common with the ability of p53 to induce the CKI $p21^{Waf1/Cip1}$ and the pro-apoptotic protein Bax, the FoxO family of transcription factors also function to induce cell cycle arrest via transcriptional upregulation of $p27^{Kip1}$ as well as GADD45b (Growth Arrest and DNA Damage 45b).[53] Indeed cell cycle arrest mediated by the induction of $p27^{Kip1}$ by FoxO family members has been observed in breast cancer, endometrial and B-cells.[54]

3. Cell Death

Cell death can be regarded as a normal process in development and homeostasis, but is also a key feature of certain disease states. For example, during embryonic development there is extensive cell death in the webbing between the digits, and in the thymus the auto-reactive T-cells undergo programmed cell death during maturation. In normal circumstances, these phenomena are required in order to shape tissue architecture and remove unwanted and potentially dangerous cells from the body. Conversely, it has been shown that the pathways that regulate cell death are often subverted in cancer, leading to an overall increase in cell number in a tumour. Consequently, the molecular mechanisms that control cell death are tightly regulated, and are highly conserved across many lower and higher eukaryotes. There are two main forms of cell death, apoptosis and necrosis.[55]

Apoptosis is associated with inversion of the plasma membrane, which exposes phophatidyl serine to the extracellular environment, cell shrinkage, nuclear condensation, DNA breakdown into large and in some cells smaller nucleosomal fragments, and finally the formation of apoptotic bodies containing the cellular fragments.[55] However, *in vivo* the final stages of apoptosis are characterised by phagocytosis of the dying cell by either nearest neighbour cells, phagocytes or macrophages. The process of necrosis is markedly different. Whilst in apoptotic cells the cell membrane remains intact, preventing the cellular contents from leaking into the extracellular environment, necrotic cells typically display a loss of membrane integrity which allows some free movement of small molecules in and out of the cell. Furthermore, necrotic cells tend to swell up rather than contract, until they burst releasing the cellular contents analogous to and over-inflated balloon bursting.[55]

3.1. Apoptosis — The role of the mitochondria and the BCL-2 family

Mitochondria is an important cellular organelle for mediating apoptosis.[56] The apoptosis-activating component of the mitochondria is cytochrome C, a key component of oxidative phosphorylation-driven ATP synthesis. Interestingly, cytochrome C is transcribed from a nuclear rather than a mitochondrial gene and the protein is translated as a precursor before being imported by the mitochondria.[57] In the mitochondria a haem moiety is added to the cytochrome C by the enzyme cytochrome C haem lyase, which makes it unable to pass through the outer mitochondrial membrane.[57] This mitochondria-dependent apoptotic pathway is also referred to as 'intrinsic' or 'cytoplasmic' apoptotic pathway, as opposed to the 'extrinsic' pathway which is mediated via death receptors, including CD95, TRAIL (TNF-related apoptosis inducing ligand), TNF (tumour necrosis factor) and TGF-beta (transforming growth factor beta) receptors.

3.2. The BCL-2 family and apoptosis

One of the key protein families that regulate mitochondria-dependent apoptosis is the BCL-2 family. This family consists of anti-apoptotic (e.g. Bcl-2 and Bcl-X_L) and pro-apoptotic (e.g. Bax, Bad, Bid and Bim).[58] All Bcl-2 family members share a BH3 (Bcl-2 homology) protein domain, but there are four other possible protein domains in the Bcl-2 family of proteins, BH1, BH2 and BH4, as well as a transmembrane domain (TM domain).[59] These different domains are important in mediating the dimerisation of the family members, and consequently the degree of the apoptotic response.[59]

Bcl-2 and Bcl-X_L function in the mitochondria of healthy cells to prevent the exit of key proteins such as cytochrome C, thereby preventing apoptosis. In normal mitochondrial metabolism, ATP exits the mitochondrial matrix via the 'adenine nucleotide translocator' (ANT) into the 'inter-mitochondrial space' (IMS), before passing through the 'voltage dependent anion channel' (VDAC) into the cytoplasm (Fig. 8).[60] Conversely, ADP travels in the reverse direction, and so a chemical equilibrium is established which regulates the concentrations of ATP and ADP in the mitochondria and the cytoplasm. ANT and VDAC are collectively known as the permeability transition pore (Fig. 8).[60,61] It has been demonstrated that cells undergo mitochondrial changes after receiving a cellular insult that induces apoptosis. Typically, these include an alteration of the balance of ATP/ADT such that ATP accumulates in the mitochondria and ADP in the cytosol.[60–62] The latter results in mitochondrial hyperpolarisation and swelling which allow the efflux of cytochrome C into the cytosol. With respect to Bcl-2 and Bcl-X_L, both of these proteins have been shown to function in the IMS to regulate the gating of both ANT and VDAC, and thus the control of ATP/ADP ration by the permeability transition pore.[62,63] Consequently, the highly regulated opening of the ANT and VDAC proteins by Bcl-2 and Bcl-X_L allows the re-establishment of the normal ADP/ATP gradient, and thus prevent apoptosis (Fig. 8).

Pro-apoptotic Bcl-2 family proteins can be sub-divided into Bax-like and BH3-only sub-groups. Bax-like proteins include Bax, Bak and Bok all containing the BH1, -2 and -3 domains. In normal cells Bax is found in the cytosol, whilst Bak is located in the mitochondrion. Following cellular insults, Bax enters the outer mitochondrial membrane and oligomerises, whilst Bak, which is already in the mitochondrion, undergoes

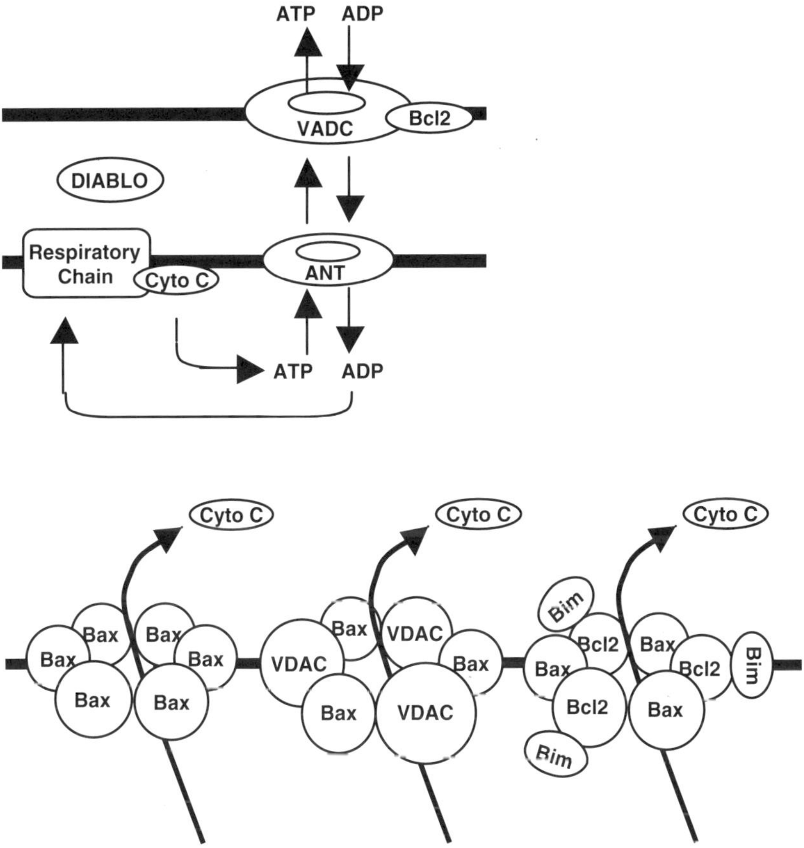

Fig. 8 Schematic detailing the membrane transport of ADP/ATP and cytochrome C. In the normal mitochondrion, ATP enters the outer mitochondrial membrane through the VDAC channel which is regulated by Bcl-2 before passing through the ANT channel in the inner mitochondrial membrane. ADP then enters the oxidative phosphorylation pathway and is converted to ATP. The ATP then exits into cytoplasm via the ANT and VDAC channels. During apoptosis, BAX is able multimerise in order to form a six-membered pore which allows the efflux of cytochrome C. Similarly, it is able to oligomerise with VDAC in order to allow cytochrome C efflux. BH3-only proteins are able to bind to Bax-associated Bcl-2 and Bcl_x in order to inhibit their anti-apoptotic function and allow the efflux of cytochrome C.

a conformational change and also oligomerises in the outer mitochondrial membrane. Furthermore, the transcription of Bax is increased following cellular insult by the tumour suppressor p53. The effect of this oligomerisation is to allow the efflux of cytochrome C and other pro-apoptotic molecules such as Diablo from the mitochondria. The precise mechanism whereby this is achieved is unclear, but one possibility is that the oligomers form large pores in the membrane, either alone or with VDAC molecules, since it has been demonstrated that Bax opens VDAC pores, whilst Bcl-X_L closes them.[60–62]

BH3-only proteins, as their name suggests only contain the BH3 domain and include Bim, Bmf, Bad, Bid, Noxa and Puma. BH3-only proteins function by binding to and inhibiting Bcl-2 and Bcl-X_L (Fig. 9).[64] The control of BH3 proteins is highly complex, with the expression of Bax, Noxa and Puma being controlled at the level of transcription by p53.[65–67] Bim is also controlled transcriptionally in part by members of the FoxO family of transcription factors, and basal levels of Bim are increased following treatment with cytotoxic drugs or cytokine withdrawal.[54,68] Additionally, in healthy cells Bim and Bmf bind to dynein light chains sequestering them to the dynein motor complex and myosin motor complex. Bad is phosphorylated by protein kinase B (also called Akt) resulting in its binding to the 14-3-3 family of phosphoprotein-binding proteins, and Bid is translated as a zymogen, requiring proteolytic activation.[64] The role of BH3 proteins is in many respects to fine-tune the apoptotic response, as BH3-only proteins are generally unable to illicit cell death on their own and require Bax or Bak. Death signals can increase Noxa and Puma levels, cause Bad, Bim and Bmf to dissociate from 14-3-3 and dynein light chain, respectively, as well as induce proteolytic activation of Bid.[64]

The regulation of apoptotic can be viewed as a balance between the anti-apoptotic function of anti-apoptotic proteins, such as Bcl-2 and Bcl-X_L, keeping cytochrome C and Diablo inside mitochondria and Bax and Bak inducing their efflux. The BH3-only proteins would appear to act as a tuning mechanism which can accelerate apoptosis by binding to and inactivating Bcl-2 and Bcl-X_L. The exit of cytochrome C and Diablo from the mitochondria initiates a cascade of protease activation that finally results in apoptosis.

3.3. Apoptosome, APAF and caspase activation

During apoptosis, the main molecules that drive the destruction of the cellular components are a family of proteins called caspases. However, caspases

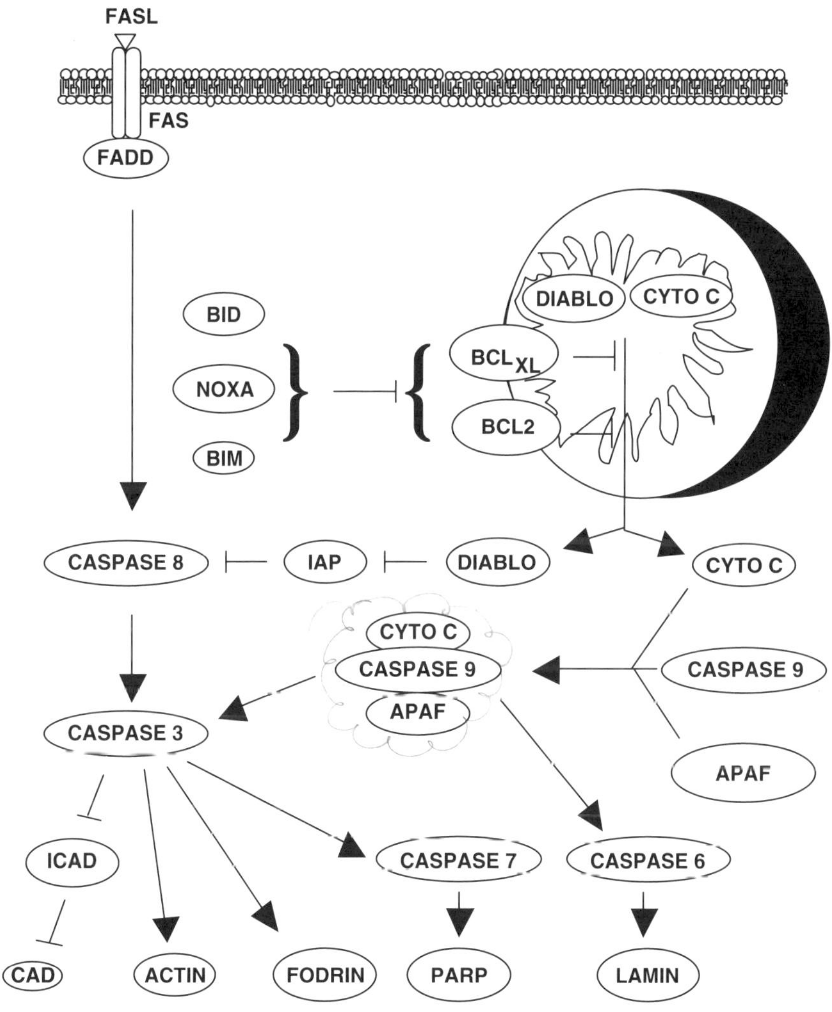

Fig. 9 General schematic detailing the extrinsic and intrinsic apoptotic pathways. Binding of FASL to the FAS receptor activates the FAS-associated FADD domain which directly activates caspase 8. Cellular stress also activates apoptosis via the mitochondrial or intrinsic pathway. In healthy cells, Bcl-2 and Bcl_X prevent the efflux of DIABLO and cytochrome C from the mitochondria whilst inhibiting the pro-apoptotic proteins such as Bax promoting the efflux of cytochrome C and DIABLO by forming pores. Apoptotic signals serve to increase the levels of these pro-apoptotic proteins as well as BH3-only proteins such as Bim, Bid and NOXA, which directly inhibit Bcl-2 and Bcl_X . Newly released cytochrome C binds to the APAF protein which then activates caspase 9. Caspases are normally inhibited by IAP proteins, however, the newly released DIABLO protein inhibits IAPs thereby relieving the blockade of caspases. Caspases 8 and 9 initiate a proteolytic cascade activating caspases 3, 7 and 6. These caspases then act on their cellular targets resulting in the breakdown of DNA into internucleosomal fragments, as well as actin, fodrin, PARP and lamin.

are synthesised as zymogen molecules, requiring proteolytic cleavage of the pro-caspase molecule in order to become biochemically active. This is achieved by a large multiprotein complex named the apoptosome.[69,70]

In mammalian cells, one of the key activators of caspase activity is APAF. APAF contains an N-terminal CARD (caspase recruitment domain), a nucleotide-binding domain, and towards the C-terminus 13 WD-40 repeats which mediate protein-protein interactions.[69,70] In healthy cells, it has been postulated that the WD40 repeats form the upper lobes of a Y-shaped monomer which bind to the CARD and part of the NBD, maintaining the protein in an inactive state. After initiation of apoptosis, cytochrome C released from the mitochondria binds to APAF monomers via the WD-40 domain which induces a conformational change giving the protein a more open configuration (Fig. 10).[69,70] The APAF then binds ATP/ADP via its NBD, and forms a wheel-like structure, comprising spokes of seven Y-shaped APAF proteins joined at the hub of the "Y" via their CARD domains.[70,71] This apoptosome complex is then able to recruit pro-caspase 9 to the hub via the CARD domains (Fig. 10). The large pro-caspase 9 undergoes a conformational change which allows the binding of a further pro-caspase 9 molecule, resulting in the formation of the active holoenzyme.[70,71] It is interesting to note that the majority of active caspase 9 is located on the apoptosome. Active caspase 9 is then able to initiate a cascade of caspase activation which ultimately results in cell death.[69–72]

3.4. Receptor activation of apoptosis

In addition to the activation of the apoptotic pathway by cellular stress, apoptosis can also be initiated by so-called death receptors. These are a class of membrane-associated receptors which share a common cysteine-rich extracellular domain and an 80-amino acid residue cytoplasmic 'death domain'. The most thoroughly characterised death receptors are Fas (also called *CD95* or APO-1), TNFα receptor (or TNFR1) and TRAIL receptor 1 and 2.[73,74] The ligands which induce apoptosis via the death receptors are FAS ligand, TNFα and Trail. In normal cells, and in the absence of ligand, death receptors do not induce apoptosis, but following the addition of ligand they cluster and can often be found in lipid rafts. This clustering of the receptors promotes an association with an adaptor protein called FADD

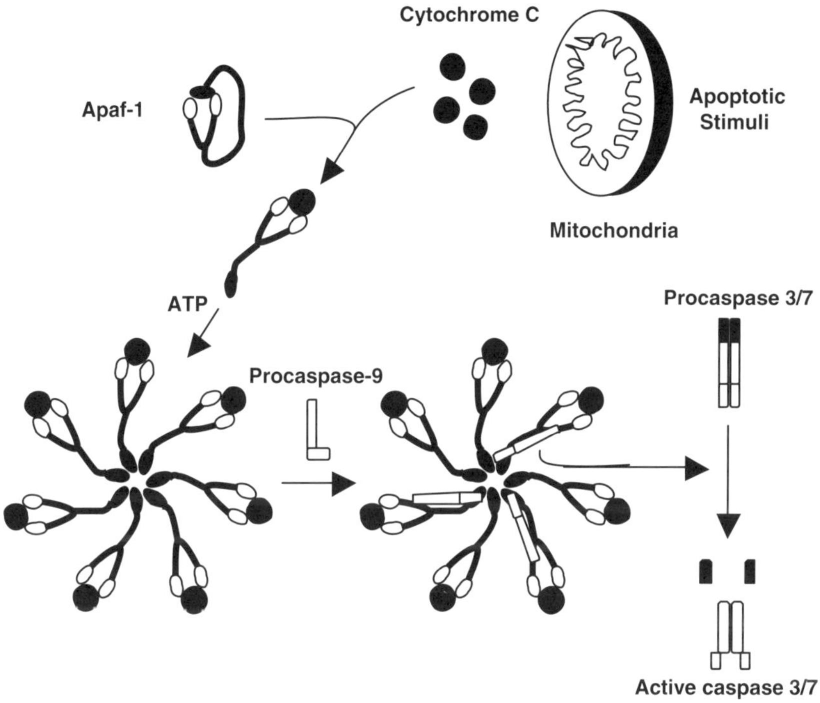

Fig. 10 Schematic representing caspase activation via APAF and the apoptosome. During apoptosis, binding of newly released cytochrome C to the APAF protein induces a conformational change in the APAF, causing its ATP-binding domain to be revealed. Binding of ATP to APAF causes it to oligomerise via the ATP-binding domain. This oligomer resembles a seven-spoked wheel and has the ability to recruit pro-caspase 9. The close proximity of bound caspase 9 molecules allows them to become active and catalyse the zymogen activation of pro-caspases 3 and 7 to their active forms.

(Fas Associated Death Domain) (Fig. 9). The receptor clustering and FADD binding then results in the binding of caspase 8 or 10 to this complex to form the DISC (Death Inducing Signalling Complex). As with the association of caspase 9 in the apoptosome, the close physical association of the newly recruited caspase 8 or 10 molecules allows their auto-proteolytic activation, whereupon active caspase molecules are released into the cytoplasm and fresh inactive caspase molecules replace them at the DISC.[74–76] Furthermore, active caspase 8 is capable of cleaving the BCL-2 family member Bid to a shorter form called tBid, which is capable of inducing mitochondrial-mediated apoptosis via cytochrome C release. Active caspase 8 or 10, just as caspase 9, are capable of activating further caspases that activate targets resulting in cell death (Fig. 9).[74–76]

3.5. The caspase cascade

Activation of the initiator caspase 9 by mitochondrial cytochrome C release, or caspase 8 and 10 by death receptor activation, initiates a cascade of caspase activation. Caspase 3 and 7 are present as dimmers within the cell, and are cleaved by the initiator caspases. The effector caspases, caspase 3, 6 and 7, are activated by proteolysis by the initiator caspases (Fig. 9). Caspase 3 substrates include those involved with cell shrinkage and membrane blebbing such as the structural proteins actin, fodrin and the protein kinase ROCK1. Other substrates include ICAD, the inhibitor of the nuclease CAD which breaks down the nuclear DNA and polyADPribosyltransferase (PARP), which is involved in protecting DNA from degradation, as well as activating proteolysis of caspase 6 and 7. Caspase 6 substrates include the nuclear envelope protein lamin A, whilst caspase 7 substrates include PARP. Consequently, the activation of caspase 3, 6 and 7 results in the destruction of the cell (Fig. 9).

There are cellular inhibitors of caspase 3, 7 and 9 called the IAP family that function to prevent uncontrolled apoptosis. In humans there are eight IAPs, the X-linked XIAP, c-IAP1, c-IAP2, melanoma-linked or ML-IAP, Livin IAP like protein 2 (ILP2), Neuronal apoptosis inhibitory protein (NAIP), Bruce and Survivin.[64,70,77] These target the various caspases in order to inhibit their proteolytic activity. IAPs contain a baculovirus like domain called a BIR domain which allows binding to caspases, and thus inhibition of apoptosis (Fig. 9).

However, the inhibition of caspases by IAPs can be abrogated by DIABLO. As mentioned previously, DIABLO is released from the mitochondrion along with cytochrome C during apoptosis.[72] DIABLO contains an IAP-like binding site to bind the BIR domains of IAPS, thereby releasing caspases from inhibition allowing apoptosis to proceed.[64,72]

3.6. Tissue specificity and cellular context

The experimental systems that have been used to identify the mechanisms of cell cycle regulation and apoptosis have mainly been based on cell and molecular biology techniques. However, data from gene knockout and transgenic mice, have shown that the activity of cell cycle and apoptotic regulatory proteins is essential to the normal function of specific cell types in development, homeostasis and disease. Due to the similarities between some key members of gene families, it is often the case that some genes are redundant, as other family members can compensate for their loss. However, whilst the animals appear to develop normally, vigorous investigations have often revealed cell and tissue specific abnormalities.

4. *In Vivo* Roles of Cell Cycle and Apoptosis Regulators

4.1. Cell cycle regulators

With respect to cyclins and CDKs, it has been shown that mice lacking cyclin D1 develop normally, but are of reduced size and have defects in the lobulo-alveolar development of the breast as well as displaying hypoplastic retinopathy.[78] Interestingly, mice over-expressing cyclin D1 develop breast and hepatocellular carcinomas.[1,79] In humans, the over-expression and in some cases amplification of the cyclin D1 gene is observed in breast, head and neck and lung cancers, additionally, in some lymphomas the cyclin D1 coding region is involved in chromosomal rearrangements leading to its aberrant expression.[80] In contrast cyclin D2 knockout mice have impaired proliferation of peripheral B-cells and ovarian granulosa formation leading to female sterility, whereas males are fertile despite hypoplastic testes.[81,82] These mice also have abnormal cerebella.[83] Cyclin D3 knockout mice are viable, but have defects in T-cell development.[84,85] Cyclin D3 is often over-expressed in thyroid cancers.[86] Mice lacking either of the cyclin E1 or E2 genes are normal, but mice lacking both die *in utero* due to defects in placentation.[82,87] In the case of A-type cyclins, mice lacking cyclin A2 die during embryonic development, whilst male mice lacking cyclin A1, which is only expressed in the germ cells, are sterile due to an inhibition of meiosis.[88] The loss of cyclin B1 is an embryonic lethal mutation, whilst the loss of cyclin B2 has no obvious defects due to compensation by cyclin B1.[89] The deletion of CDK4 results in viable animals, but of diminished size with both male and female sterility, and due to defects in the β-islet cells develop type 2 diabetes.[82] Whilst the keratinocyte differentiation pathway is unaffected, there is a reduction in skin tumourigenesis.[90] By contrast mice over-expressing CDK4 develop melanomas,[91] and CDK4 expression is often elevated in bone cancers.[92] Mice lacking CDK6 display delayed entry into S-phase in their T-cells. The loss of CDK2 leads to a failure of meiosis and sterility in both sexes.[82]

The loss of the CKI $p21^{Waf1/Cip1}$ in mice causes no physiological abnormalities, but cells are defective in the G_1 checkpoint control[93] and show an increased susceptibility to carcinogenesis.[94] It has also been demonstrated that $p21^{Waf1/Cip1}$ is important in muscle cell differentiation,[95] but that the loss of $p21^{Waf1/Cip1}$ is compensated for by other CKIs. The knockout of the $p27^{Kip1}$ locus results in mice which are significantly larger than normal, which display abnormalities in the pituitary gland and ovaries, as well as in the bone-forming osteoblast cell population.[96–99] The loss of $p27^{Kip1}$ expression has been observed in many human cancers especially breast cancer.[100] Mutations in the $p57^{Kip2}$ locus have been associated with the human disease Beckwith-Wiedemann syndrome, a condition which is characterised by macro glossia (large tongues) and cancer pre-disposition,[101] whilst mice lacking $p57^{Kip2}$ show some disorganisation in the structure of their long bones due to chondrocyte abnormalities.[102] The loss of $p57^{Kip2}$ expression has also been implicated in lung cancer.[103] The loss of the retinoblastoma protein Rb is lethal, but in humans who have lost only one copy of the gene are prone to the rare eye tumour retinoblastoma, as well as osteosarcoma in later life, whilst the corresponding mice develop pituitary tumours.[104] Mutations in the p16 locus have also been described in human melanoma, but these have been ascribed to artifacts of tissue culture.[105] Furthermore, the *ink4a* locus also generates an additional transcript with an alternate reading frame called 14 or p19 ARF (in humans and mice, respectively) which is involved in the p53 MDM2/hMDM pathway.[106] However, loss of the whole locus in mice results in increased tumour formation of fibrosarcomas and B-cell lymphomas, as well as extramedullary haematopoiesis. Loss of either of the other family members, p107 and p130 in mice does not result in any major phenotype, but loss of both p107 and p130 causes mice to develop with abnormal long bone due to deregulated chondrocyte growth.[107]

In general terms, these phenotypes reflect the functions of the proteins involved, i.e. loss of CKIs can result in larger animals, and loss of cyclins and CDKs result in smaller animals and stunted cell populations as a result of a reduction in proliferation.

4.2. Apoptotic regulators

As with the loss of cell cycle regulators, the loss of proteins that control apoptosis in experimental animals results in a series of phenotypes which highlight the specific functions of key players in the regulation of apoptosis. Knockout mice lacking *Bcl-2* are smaller, die within months of birth, display craniofacial abnormalities and are subject to renal failure as well as excessive apoptosis in the thymus and spleen.[108] Bcl-2 over-expression has also been observed in many human tumours where it is thought to prevent cell death and thus increase the population of transformed cells. *Bcl-x* knockout mice also display reduced apoptosis of haematopoeitic cells, as well as neurones and die during embryogenesis.[109] Of the pro-apoptotic regulators, loss of the *Bax* allele results in viable animals with mild lymphoid hyperplasia, male infertility and extended oocyte lifespan.[110] Mice deficient in the Bak gene have no obvious phenotype, but loss of both *Bax* and *Bak* results in perinatal lethality, with splenomegaly, an accumulation of neurones in the brain and a persistence of the interdigital webbing.[111] The loss of the BH3-only protein Bim causes partial embryonic lethality, with surviving animals displaying splenomegaly, as well as increased plasma cell dysplasia and premature death due to glomerulonephritis, cardiac infarction and vasculitis.[112] The loss of Bid only results in the Fas-mediated apoptosis in the liver.[113] Of the inducers of apoptosis, the loss of APAF1 causes embryonic lethality, with a persistent interdigital web, defective neuronal apoptosis which results in exencephaly and heart defects.[114,115] Of the caspases, the loss of caspase 3 results in perinatal lethality with neuronal hyperplasia.[116] Loss of caspase 3 activity has also been observed in some breast cancer cell lines.[117] The loss of caspase 8 results in embryonic lethality, with resistance to death receptor-induced apoptosis.[118] Caspase 9 knockout mice develop perinatal lethality with heart defects.[119]

These results show that the loss of pro-apoptotic proteins often results in increased cell number whilst the loss of anti-apoptotic proteins resulted in increased apoptosis. The more obvious effects of changes in the cell cycle and apoptotic regulators, i.e. increased growth in cells lacking CKIs or pro-apoptotic proteins, or over-expressing cyclins or anti-apoptotic proteins would arguably be predictable from their cellular function. However, the phenotypes that do arise is some animals, for example the breast defect in cyclin D1-deficient mice, suggests that there are some unique functions for these proteins in specific tissues. How these proteins function in these specific situations is as yet unclear.

5. The Role of Phosphatidylinositol 3-Kinase in Proliferation and Apoptosis

One key signal transduction pathway that regulates cell cycle progression and apoptosis is the phosphatidylinositol 3-kinase (PI3K) pathway,[120] which is stimulated by a number of growth factors.[120] The role of the PI3K pathway in proliferation and apoptosis will be reviewed briefly here, but the reader is referred to the other chapters within this edition for a fuller explanation of the PI3K pathway and its implication. Once a receptor tyrosine kinase has become activated by binding a specific ligand, it becomes autophosphorylated and binds PI3K heterodimers either directly or indirectly via IRS adaptor proteins.[120–122] The interaction between the p85 regulatory subunit of PI3K and the receptor brings the catalytic domain (p110 subunit) of PI3K into contact with the plasma membrane, where it is able to phosphorylate its targets. Alternatively, PI3K can be juxtaposed to the plasma membrane via interactions with activated RAS family members. PI3K phosphorylates phosphotydylinositol (4,5) diphosphate (PIP_2) at the 3 position, resulting in the formation of phosphotydylinositol (3,4,5) triphosphate (PIP_3). PIP_3 binds to protein kinase B (PKB) also called Akt, a serine/threonine kinase, via its pleckstrin homology (PH) domain and recruits it to the inner surface of the cell membrane where it becomes activated by another PH domain-containing protein, PDK1.[120–122] PKB phosphorylates a number of key substrates including glycogen synthase kinase 3β (GSKβ) and mTOR, which is involved in cell growth and translation, and the transcription factor NFκB.[120]

The PI3K promotes cell cycle progression through a number of ways. Firstly, Akt-mediated inhibition of GSKβ which relieves the inhibition of β catenin, resulting in increased LEF-mediated transcription of the cyclin D1 gene.[123] Furthermore, GSKβ phosphorylation of cyclin D1 results in its cytoplasmic localisation and proteolysis, so inhibition of GSKβ acts to increase the protein levels of cyclin D1 as well.[124] The FOXO family of transcription factors which are regulated by Akt have also been shown to regulate cell cycle progression. FOXOs are responsible for the transcriptional inhibition of cyclins D1 and D2 and their inhibition by Akt results in increased rates of transcription.[125,126] Furthermore, FOXO3a has been shown in different cell types to increase the transcription of the $p27^{Kip1}$ gene.[127] Additionally, Akt is capable of phosphorylating both $p27^{Kip1}$ and $p21^{Waf1/Cip1}$, thereby inhibiting their anti-proliferative activity.[128,129]

The role for PI3K-Akt signalling in apoptosis is also important. The pro-apoptotic protein BAD is phosphorylated and inhibited by Akt.[130,131] Inhibition of FOXOs by Akt results in repression of the transcription of the pro-apoptotic genes Bim[54,68,132] and FASL.[133] Furthermore, the phosphorylation of cAMP-response element binding proteins promotes the transcription of the anti-apoptotic proteins Bcl-2 and Mcl-1.[134,135] Consequently, the role of the PI3K-Akt pathway is that of a nexus through which various signalling molecules act in order to promote changes in the rates of cell growth and death. Interaction of the PI3K pathway with other pathways modulates these effects to produce the tissue-specific effects needed for the normal homeostasis observed during development and adulthood.

6. Summary

The mechanisms outlined here deal with the key regulators of cell cycle regulation and apoptosis, however these pathways must also be viewed in context. The various biological processes that regulate normal development and homeostasis also directly regulate cell division and apoptosis in order to create an organism with the correct tissue architecture and function. What is also clear is that these pathways can be subject to aberrant regulation during disease, most notably in cancer, where oncogenes are often responsible for increased growth rates and decreased levels of apoptosis.

References

1. C.J. Sherr, Cancer cell cycles, *Science* **274** (1996) 1672–1677.
2. M. Sugimoto, N. Martin, D.P. Wilks, K. Tamai, T.J. Huot, C. Pantoja, K. Okumura, M. Serrano and E. Hara, Activation of cyclin D1-kinase in

murine fibroblasts lacking both p21(Cip1) and p27(Kip1), *Oncogene* **21** (2002) 8067–8074.
3. J. LaBaer, M.D. Garrett, L.F. Stevenson, J.M. Slingerland, C. Sandhu, H.S. Chou, A. Fattaey and E. Harlow, New functional activities for the p21 family of CDK inhibitors, *Genes Dev.* **11** (1997) 847–862.
4. J.N. Lavoie, G. L'Allemain, A. Brunet, R. Muller and J. Pouyssegur, Cyclin D1 expression is regulated positively by the p42/p44MAPK and negatively by the p38/HOGMAPK pathway, *J. Biol. Chem.* **271** (1996) 20608–20616.
5. J. Glassford, I. Soeiro, S.M. Skarell, L. Banerji, M. Holman, G.G. Klaus, T. Kadowaki, S. Koyasu and E.W.-F. Lam, BCR targets cyclin D2 via Btk and the p85alpha subunit of PI3K to induce cell cycle progression in primary mouse B cells, *Oncogene* **22** (2003) 2248–2259.
6. V. Baldin, J. Lukas, M.J. Marcote, M. Pagano and G. Draetta, Cyclin D1 is a nuclear protein required for cell cycle progression in G1, *Genes Dev.* **7** (1993) 812–821.
7. J. Bartek, J. Bartkova and J. Lukas, The retinoblastoma protein pathway and the restriction point, *Curr. Opin. Cell Biol.* **8** (1996) 805–814.
8. S.P. Chellappan, S. Hiebert, M. Mudryj, J.M. Horowitz and J.R. Nevins, The E2F transcription factor is a cellular target for the RB protein, *Cell* **65** (1991) 1053–1061.
9. N. Dyson, pRB, p107 and the regulation of the E2F transcription factor, *J. Cell Sci. Suppl.* **18** (1994) 81–87.
10. A. Brehm, E.A. Miska, D.J. McCance, J.L. Reid, A.J. Bannister and T. Kouzarides, Retinoblastoma protein recruits histone deacetylase to repress transcription, *Nature* **391** (1998) 597–601 (see comments).
11. K. Helin, E. Harlow and A. Fattaey, Inhibition of E2F-1 transactivation by direct binding of the retinoblastoma protein, *Mol. Cell. Biol.* **13** (1993) 6501–6508.
12. E.S. Knudsen and J.Y. Wang, Dual mechanisms for the inhibition of E2F binding to RB by cyclin-dependent kinase-mediated RB phosphorylation, *Mol. Cell. Biol.* **17** (1997) 5771–5783.
13. J. Kato, H. Matsushime, S.W. Hiebert, M.E. Ewen and C.J. Sherr, Direct binding of cyclin D to the retinoblastoma gene product (pRb) and pRb phosphorylation by the cyclin D-dependent kinase CDK4, *Genes Dev.* **7** (1993) 331–342.
14. G. Peters, The D-type cyclins and their role in tumorigenesis, *J. Cell Sci. Suppl.* **18** (1994) 89–96.

15. L. Connell-Crowley, J.W. Harper and D.W. Goodrich, Cyclin D1/Cdk4 regulates retinoblastoma protein-mediated cell cycle arrest by site-specific phosphorylation, *Mol. Biol. Cell* **8** (1997) 287–301.
16. J.W. Harbour, R.X. Luo, A. Dei Santi, A.A. Postigo and D.C. Dean, Cdk phosphorylation triggers sequential intramolecular interactions that progressively block Rb functions as cells move through G1, *Cell* **98** (1999) 859–869.
17. K. Ohtani, J. DeGregori and J.R. Nevins, Regulation of the cyclin E gene by transcription factor E2F1, *Proc. Natl. Acad. Sci. USA* **92** (1995) 12146–12150.
18. Y. Geng, E.N. Eaton, M. Picon, J.M. Roberts, A.S. Lundberg, A. Gifford, C. Sardet and R.A. Weinberg, Regulation of cyclin E transcription by E2Fs and retinoblastoma protein, *Oncogene* **12** (1996) 1173–1180.
19. C.J. Sherr, The ins and outs of RB: coupling gene expression to the cell cycle clock, *Trends Cell Biol.* **4** (1994) 15–18.
20. C.J. Sherr, The Pezcoller lecture: cancer cell cycles revisited, *Cancer Res.* **60** (2000) 3689–3695.
21. J. Fan and J.R. Bertino, Functional roles of E2F in cell cycle regulation, *Oncogene* **14** (1997) 1191–1200.
22. A. Early, L.S. Drury and J.F. Diffley, Mechanisms involved in regulating DNA replication origins during the cell cycle and in response to DNA damage, *Philos. Trans. R. Soc. Lond. B. Biol. Sci.* **359** (2004) 31–38.
23. H. Nishitani and Z. Lygerou, Control of DNA replication licensing in a cell cycle, *Genes Cells* **7** (2002) 523–534.
24. M. Findeisen, M. El-Denary, T. Kapitza, R. Graf and U. Strausfeld, Cyclin A-dependent kinase activity affects chromatin binding of ORC, Cdc6, and MCM in egg extracts of *Xenopus laevis, Eur. J. Biochem.* **264** (1999) 415–426.
25. E.E. Arias and J.C. Walter, Replication-dependent destruction of Cdt1 limits DNA replication to a single round per cell cycle in Xenopus egg extracts, *Genes Dev.* **19** (2005) 114–126.
26. L.L. Parker and H. Piwnica-Worms, Inactivation of the p34cdc2-cyclin B complex by the human WEE1 tyrosine kinase, *Science* **257** (1992) 1955–1957.
27. R.N. Booher, P.S. Holman and A. Fattaey, Human Myt1 is a cell cycle-regulated kinase that inhibits Cdc2 but not Cdk2 activity, *J. Biol. Chem.* **272** (1997) 22300–22306.
28. D.O. Morgan, Principles of CDK regulation, *Nature* **374** (1995) 131–134.

29. L.A. Porter and D.J. Donoghue, Cyclin B1 and CDK1: nuclear localization and upstream regulators, *Prog. Cell Cycle Res.* **5** (2003) 335–347.
30. I. Hoffmann, P.R. Clarke, M.J. Marcote, E. Karsenti and G. Draetta, Phosphorylation and activation of human cdc25-C by cdc2-cyclin B and its involvement in the self-amplification of MPF at mitosis, *EMBO J.* **12** (1993) 53–63.
31. A.K. Roshak, E.A. Capper, C. Imburgia, J. Fornwald, G. Scott and L.A. Marshall, The human polo-like kinase, PLK, regulates cdc2/cyclin B through phosphorylation and activation of the cdc25C phosphatase, *Cell. Signal.* **12** (2000) 405–411.
32. C.Y. Peng, P.R. Graves, R.S. Thoma, Z. Wu, A.S. Shaw and H. Piwnica-Worms, Mitotic and G_2 checkpoint control: regulation of 14-3-3 protein binding by phosphorylation of Cdc25C on serine-216, *Science* **277** (1997) 1501–1505.
33. E.A. Nigg, Cellular substrates of p34(cdc2) and its companion cyclin-dependent kinases, *Trends Cell Biol.* **3** (1993) 296–301.
34. V.A. Smits and R.H. Medema, Checking out the G(2)/M transition, *Biochim. Biophys. Acta* **1519** (2001) 1–12.
35. E.A. Nigg, Mitotic kinases as regulators of cell division and its checkpoints, *Nat. Rev. Mol. Cell Biol.* **2** (2001) 21–32.
36. T. Mayor, Y.D. Stierhof, K. Tanaka, A.M. Fry and E.A. Nigg, The centrosomal protein C-Nap1 is required for cell cycle-regulated centrosome cohesion, *J. Cell Biol.* **151** (2000) 837–846.
37. T. Mayor, U. Hacker, Y.D. Stierhof and E.A. Nigg, The mechanism regulating the dissociation of the centrosomal protein C-Nap1 from mitotic spindle poles, *J. Cell Sci.* **115** (2002) 3275–3284.
38. J.M. Peters, The anaphase-promoting complex: proteolysis in mitosis and beyond, *Mol. Cell* **9** (2002) 931–943.
39. F. Uhlmann, The mechanism of sister chromatid cohesion, *Exp. Cell Res.* **296** (2004) 80–85.
40. C.J. Bakkenist and M.B. Kastan, DNA damage activates ATM through intermolecular autophosphorylation and dimer dissociation, *Nature* **421** (2003) 499–506.
41. R. Kitagawa, C.J. Bakkenist, P.J. McKinnon and M.B. Kastan, Phosphorylation of SMC1 is a critical downstream event in the ATM-NBS1-BRCA1 pathway, *Genes Dev.* **18** (2004) 1423–1438.
42. J. Bartek and J. Lukas, Chk1 and Chk2 kinases in checkpoint control and cancer, *Cancer Cell* **3** (2003) 421–429.
43. R. Maya, M. Balass, S.T. Kim, D. Shkedy, J.F. Leal, O. Shifman, M. Moas, T. Buschmann, Z. Ronai, Y. Shiloh, M.B. Kastan, E. Katzir

and M. Oren, ATM-dependent phosphorylation of Mdm2 on serine 395: role in p53 activation by DNA damage, *Genes Dev.* **15** (2001) 1067–1077.
44. Y. Haupt, R. Maya, A. Kazaz and M. Oren, Mdm2 promotes the rapid degradation of p53, *Nature* **387** (1997) 296–299.
45. W.S. el-Deiry, J.W. Harper, P.M. O'Connor, V.E. Velculescu, C.E. Canman, J. Jackman, J.A. Pietenpol, M. Burrell, D.E. Hill, Y. Wang, *et al.*, WAF1/CIP1 is induced in p53-mediated G1 arrest and apoptosis, *Cancer Res.* **54** (1994) 1169–1174.
46. S. St Clair, L. Giono, S. Varmeh-Ziaie, L. Resnick-Silverman, W.J. Liu, A. Padi, J. Dastidar, A. DaCosta, M. Mattia and J.J. Manfredi, DNA damage-induced downregulation of Cdc25C is mediated by p53 via two independent mechanisms: one involves direct binding to the cdc25C promoter, *Mol. Cell* **16** (2004) 725–736.
47. H.L. Ball and D. Cortez, ATRIP oligomerization is required for ATR-dependent checkpoint signaling, *J. Biol. Chem.* **280** (2005) 31390–31396.
48. M.J. O'Connell and K.A. Cimprich, G_2 damage checkpoints: what is the turn-on? *J. Cell Sci.* **118** (2005) 1–6.
49. A. Sancar, L.A. Lindsey-Boltz, K. Unsal-Kacmaz and S. Linn, Molecular mechanisms of mammalian DNA repair and the DNA damage checkpoints, *Annu. Rev. Biochem.* **73** (2004) 39–85.
50. M.L. Agarwal, A. Agarwal, W.R. Taylor and G.R. Stark, p53 controls both the G_2/M and the G1 cell cycle checkpoints and mediates reversible growth arrest in human fibroblasts, *Proc. Natl. Acad. Sci. USA* **92** (1995) 8493–8497.
51. R.T. Abraham, Cell cycle checkpoint signaling through the ATM and ATR kinases, *Genes Dev.* **15** (2001) 2177–2196.
52. K.A. Nyberg, R.J. Michelson, C.W. Putnam and T.A. Weinert, Toward maintaining the genome: DNA damage and replication checkpoints, *Annu. Rev. Genet.* **36** (2002) 617–656.
53. H. You and T.W. Mak, Crosstalk between p53 and FOXO Transcription Factors, *Cell Cycle* **4** (2005) 37–38.
54. A. Sunters, S. Fernandez de Mattos, M. Stahl, J.J. Brosens, G. Zoumpoulidou, C.A. Saunders, P.J. Coffer, R.H. Medema, R.C. Coombes and E.W. Lam, FoxO3a transcriptional regulation of Bim controls apoptosis in paclitaxel-treated breast cancer cell lines, *J. Biol. Chem.* **278** (2003) 49795–49805.
55. W. Fiers, R. Beyaert, W. Declercq and P. Vandenabeele, More than one way to die: apoptosis, necrosis and reactive oxygen damage, *Oncogene* **18** (1999) 7719–7730.

56. D.D. Newmeyer, D.M. Farschon and J.C. Reed, Cell-free apoptosis in Xenopus egg extracts: inhibition by Bcl-2 and requirement for an organelle fraction enriched in mitochondria, *Cell* **79** (1994) 353–364.
57. A. Mayer, W. Neupert and R. Lill, Translocation of apocytochrome *c* across the outer membrane of mitochondria, *J. Biol. Chem.* **270** (1995) 12390–12397.
58. S. Cory and J.M. Adams, The Bcl-2 family: regulators of the cellular life-or-death switch, *Nat. Rev. Cancer* **2** (2002) 647–656.
59. J.C. Reed, Bcl-2 family proteins, *Oncogene* **17** (1998) 3225–3236.
60. J.M. Adams and S. Cory, Life-or-death decisions by the Bcl-2 protein family, *Trends Biochem. Sci.* **26** (2001) 61–66.
61. M.H. Harris and C.B. Thompson, The role of the Bcl-2 family in the regulation of outer mitochondrial membrane permeability, *Cell Death Differ.* **7** (2000) 1182–1191.
62. M.G. Vander Heiden and C.B. Thompson, Bcl-2 proteins: regulators of apoptosis or of mitochondrial homeostasis? *Nat. Cell Biol.* **1** (1999) E209–E216.
63. S. Shimizu, M. Narita and Y. Tsujimoto, Bcl-2 family proteins regulate the release of apoptogenic cytochrome *c* by the mitochondrial channel VDAC, *Nature* **399** (1999) 483–487.
64. N.N. Danial and S.J. Korsmeyer, Cell death: critical control points, *Cell* **116** (2004) 205–219.
65. K. Nakano and K.H. Vousden, PUMA, a novel proapoptotic gene, is induced by p53, *Mol. Cell* **7** (2001) 683–694.
66. T. Shibue, K. Takeda, E. Oda, H. Tanaka, H. Murasawa, A. Takaoka, Y. Morishita, S. Akira, T. Taniguchi and N. Tanaka, Integral role of Noxa in p53-mediated apoptotic response, *Genes Dev.* **17** (2003) 2233–2238.
67. E. Oda, R. Ohki, H. Murasawa, J. Nemoto, T. Shibue, T. Yamashita, T. Tokino, T. Taniguchi and N. Tanaka, Noxa, a BH3-only member of the Bcl-2 family and candidate mediator of p53-induced apoptosis, *Science* **288** (2000) 1053–1058.
68. P.F. Dijkers, R.H. Medemadagger, J.W. Lammers, L. Koenderman and P.J. Coffer, Expression of the pro-apoptotic Bcl-2 family member Bim is regulated by the forkhead transcription factor FKHR-L1, *Curr. Biol.* **10** (2000) 1201–1204.
69. M. Chen and J. Wang, Initiator caspases in apoptosis signaling pathways, *Apoptosis* **7** (2002) 313–319.
70. Y. Shi, Mechanisms of caspase activation and inhibition during apoptosis, *Mol. Cell* **9** (2002) 459–470.

71. S.J. Riedl and Y. Shi, Molecular mechanisms of caspase regulation during apoptosis, *Nat. Rev. Mol. Cell Biol.* **5** (2004) 897–907.
72. E.N. Shiozaki and Y. Shi, Caspases, IAPs and Smac/DIABLO: mechanisms from structural biology, *Trends Biochem. Sci.* **29** (2004) 486–494.
73. H. Wajant, Death receptors, *Essays Biochem.* **39** (2003) 53–71.
74. H. Wajant, The Fas signaling pathway: more than a paradigm, *Science* **296** (2002) 1635–1636.
75. E.F. Wagner, M. Hleb, N. Hanna and S. Sharma, A pivotal role of cyclin D3 and cyclin-dependent kinase inhibitor p27 in the regulation of IL-2-, IL-4-, or IL-10-mediated human B cell proliferation, *J. Immunol.* **161** (1998) 1123–1131.
76. U. Sartorius, I. Schmitz and P.H. Krammer, Molecular mechanisms of death-receptor-mediated apoptosis, *Chembiochem* **2** (2001) 20–29.
77. Y. Shi, Apoptosome: the cellular engine for the activation of caspase-9, *Structure (Camb)* **10** (2002) 285–288.
78. V. Fantl, G. Stamp, A. Andrews, I. Rosewell and C. Dickson, Mice lacking cyclin D1 are small and show defects in eye and mammary gland development, *Genes Dev.* **9** (1995) 2364–2372.
79. N.G. Deane, M.A. Parker, R. Aramandla, L. Diehl, W.J. Lee, M.K. Washington, L.B. Nanney, Y. Shyr and R.D. Beauchamp, Hepatocellular carcinoma results from chronic cyclin D1 overexpression in transgenic mice, *Cancer Res.* **61** (2001) 5389–5395.
80. M. Fu, C. Wang, Z. Li, T. Sakamaki and R.G. Pestell, Minireview: Cyclin D1: Normal and Abnormal Functions, *Endocrinology* **145** (2004) 5439–5447.
81. P. Sicinski, J.L. Donaher, Y. Geng, S.B. Parker, H. Gardner, M.Y. Park, R.L. Robker, J.S. Richards, L.K. McGinnis, J.D. Biggers, J.J. Eppig, R.T. Bronson, S.J. Elledge and R.A. Weinberg, Cyclin D2 is an FSH-responsive gene involved in gonadal cell proliferation and oncogenesis, *Nature* **384** (1996) 470–474.
82. C.J. Sherr and J.M. Roberts, Living with or without cyclins and cyclin-dependent kinases, *Genes Dev.* **18** (2004) 2699–2711.
83. J.M. Huard, C.C. Forster, M.L. Carter, P. Sicinski and M.E. Ross, Cerebellar histogenesis is disturbed in mice lacking cyclin D2, *Development* **126** (1999) 1927–1935.
84. J.A. Diehl, The cyclin D3 knockout: a pound of redundancy with a dash of tissue specificity, *Cancer Biol. Ther.* **3** (2004) 162–164.

85. E. Sicinska, I. Aifantis, L. Le Cam, W. Swat, C. Borowski, Q. Yu, A.A. Ferrando, S.D. Levin, Y. Geng, H. von Boehmer and P. Sicinski, Requirement for cyclin D3 in lymphocyte development and T cell leukemias, *Cancer Cell* **4** (2003) 451–461.
86. G. Baldassarre, B. Belletti, P. Bruni, A. Boccia, F. Trapasso, F. Pentimalli, M.V. Barone, G. Chiappetta, M.T. Vento, S. Spiezia, A. Fusco and G. Viglietto, Overexpressed cyclin D3 contributes to retaining the growth inhibitor p27 in the cytoplasm of thyroid tumor cells, *J. Clin. Invest.* **104** (1999) 865–874.
87. Y. Geng, Q. Yu, E. Sicinska, M. Das, J.E. Schneider, S. Bhattacharya, W.M. Rideout, R.T. Bronson, H. Gardner and P. Sicinski, Cyclin E ablation in the mouse, *Cell* **114** (2003) 431–443.
88. D. Liu, M.M. Matzuk, W.K. Sung, Q. Guo, P. Wang and D.J. Wolgemuth, Cyclin A1 is required for meiosis in the male mouse, *Nat. Genet.* **20** (1998) 377–380.
89. M. Brandeis, I. Rosewell, M. Carrington, T. Crompton, M.A. Jacobs, J. Kirk, J. Gannon and T. Hunt, Cyclin B2-null mice develop normally and are fertile whereas cyclin B1-null mice die in utero, *Proc. Natl. Acad. Sci. USA* **95** (1998) 4344–4349.
90. M.L. Rodriguez-Puebla, P.L. Miliani de Marval, M. LaCava, D.S. Moons, H. Kiyokawa and C.J. Conti, Cdk4 deficiency inhibits skin tumor development but does not affect normal keratinocyte proliferation, *Am. J. Pathol.* **161** (2002) 405–411.
91. R. Sotillo, J.F. Garcia, S. Ortega, J. Martin, P. Dubus, M. Barbacid and M. Malumbres, Invasive melanoma in Cdk4-targeted mice, *Proc. Natl. Acad. Sci. USA* **98** (2001) 13312–13317.
92. M.S. Benassi, L. Molendini, G. Gamberi, M.R. Sollazzo, P. Ragazzini, M. Merli, G. Magagnoli, L. Sangiorgi, P. Bacchini, F. Bertoni and P. Picci, Altered G1 phase regulation in osteosarcoma, *Int. J. Cancer* **74** (1997) 518–522.
93. C. Deng, P. Zhang, J.W. Harper, S.J. Elledge and P. Leder, Mice lacking p21CIP1/WAF1 undergo normal development, but are defective in G1 checkpoint control, *Cell* **82** (1995) 675–684.
94. J. Martin-Caballero, J.M. Flores, P. Garcia-Palencia and M. Serrano, Tumor susceptibility of p21(Waf1/Cip1)-deficient mice, *Cancer Res.* **61** (2001) 6234–6238.
95. O. Halevy, B.G. Novitch, D.B. Spicer, S.X. Skapek, J. Rhee, G.J. Hannon, D. Beach and A.B. Lassar, Correlation of terminal cell cycle arrest of skeletal muscle with induction of p21 by MyoD, *Science* **267** (1995) 1018–1021.

96. K. Nakayama, N. Ishida, M. Shirane, A. Inomata, T. Inoue, N. Shishido, I. Horii and D.Y. Loh, Mice lacking p27(Kip1) display increased body size, multiple organ hyperplasia, retinal dysplasia, and pituitary tumors, *Cell* **85** (1996) 707–720.
97. H. Kiyokawa, R.D. Kineman, K.O. Manova-Todorova, V.C. Soares, E.S. Hoffman, M. Ono, D. Khanam, A.C. Hayday, L.A. Frohman and A. Koff, Enhanced growth of mice lacking the cyclin-dependent kinase inhibitor function of p27(Kip1), *Cell* **85** (1996) 721–732.
98. M.L. Fero, M. Rivkin, M. Tasch, P. Porter, C.E. Carow, E. Firpo, K. Polyak, L.H. Tsai, V. Broudy, R.M. Perlmutter, K. Kaushansky and J.M. Roberts, A syndrome of multiorgan hyperplasia with features of gigantism, tumorigenesis, and female sterility in p27(Kip1)-deficient mice, *Cell* **85** (1996) 733–744.
99. H. Drissi, D. Hushka, F. Aslam, Q. Nguyen, E. Buffone, A. Koff, A. van Wijnen, J.B. Lian, J.L. Stein and G.S. Stein, The cell cycle regulator p27kip1 contributes to growth and differentiation of osteoblasts, *Cancer Res.* **59** (1999) 3705–3711.
100. R. Chiarle, M. Pagano and G. Inghirami, The cyclin dependent kinase inhibitor p27 and its prognostic role in breast cancer, *Breast Cancer Res.* **3** (2001) 91–94.
101. W.J. Swanger and J.M. Roberts, p57KIP2 targeted disruption and Beckwith-Wiedemann syndrome: is the inhibitor just a contributor? *Bioessays* **19** (1997) 839–842.
102. Y. Yan, J. Frisen, M.H. Lee, J. Massague and M. Barbacid, Ablation of the CDK inhibitor p57Kip2 results in increased apoptosis and delayed differentiation during mouse development, *Genes Dev.* **11** (1997) 973–983.
103. M. Kondo, S. Matsuoka, K. Uchida, H. Osada, M. Nagatake, K. Takagi, J.W. Harper, T. Takahashi and S.J. Elledge, Selective maternal-allele loss in human lung cancers of the maternally expressed p57KIP2 gene at 11p15.5, *Oncogene* **12** (1996) 1365–1368.
104. M. Vooijs and A. Berns, Developmental defects and tumor predisposition in Rb mutant mice, *Oncogene* **18** (1999) 5293–5303.
105. M. Serrano, The tumor suppressor protein p16INK4a, *Exp. Cell Res.* **237** (1997) 7–13.
106. D.A. Haber, Splicing into senescence: the curious case of p16 and p19ARF, *Cell* **91** (1997) 555–558.
107. D. Cobrinik, M.H. Lee, G. Hannon, G. Mulligan, R.T. Bronson, N. Dyson, E. Harlow, D. Beach, R.A. Weinberg and T. Jacks, Shared role of the pRB-related p130 and p107 proteins in limb development, *Genes Dev.* **10** (1996) 1633–1644.

108. D.J. Veis, C.M. Sorenson, J.R. Shutter and S.J. Korsmeyer, Bcl-2-deficient mice demonstrate fulminant lymphoid apoptosis, polycystic kidneys, and hypopigmented hair, *Cell* **75** (1993) 229–240.
109. N. Motoyama, F. Wang, K.A. Roth, H. Sawa, K. Nakayama, I. Negishi, S. Senju, Q. Zhang, S. Fujii, *et al.*, Massive cell death of immature hematopoietic cells and neurons in Bcl-x-deficient mice, *Science* **267** (1995) 1506–1510.
110. C.M. Knudson, K.S. Tung, W.G. Tourtellotte, G.A. Brown and S.J. Korsmeyer, Bax-deficient mice with lymphoid hyperplasia and male germ cell death, *Science* **270** (1995) 96–99.
111. T. Lindsten, A.J. Ross, A. King, W.X. Zong, J.C. Rathmell, H.A. Shiels, E. Ulrich, K.G. Waymire, P. Mahar, K. Frauwirth, Y. Chen, M. Wei, V.M. Eng, D.M. Adelman, M.C. Simon, A. Ma, J.A. Golden, G. Evan, S.J. Korsmeyer, G.R. MacGregor and C.B. Thompson, The combined functions of proapoptotic Bcl-2 family members bak and bax are essential for normal development of multiple tissues, *Mol. Cell* **6** (2000) 1389–1399.
112. P. Bouillet, D. Metcalf, D.C. Huang, D.M. Tarlinton, T.W. Kay, F. Kontgen, J.M. Adams and A. Strasser, Proapoptotic Bcl-2 relative Bim required for certain apoptotic responses, leukocyte homeostasis, and to preclude autoimmunity, *Science* **286** (1999) 1735–1738.
113. X.M. Yin, K. Wang, A. Gross, Y. Zhao, S. Zinkel, B. Klocke, K.A. Roth and S.J. Korsmeyer, Bid-deficient mice are resistant to Fas-induced hepatocellular apoptosis, *Nature* **400** (1999) 886–891.
114. F. Cecconi, G. Alvarez-Bolado, B.I. Meyer, K.A. Roth and P. Gruss, Apaf1 (CED-4 homolog) regulates programmed cell death in mammalian development, *Cell* **94** (1998) 727–737.
115. H. Yoshida, Y.Y. Kong, R. Yoshida, A.J. Elia, A. Hakem, R. Hakem, J.M. Penninger and T.W. Mak, Apaf1 is required for mitochondrial pathways of apoptosis and brain development, *Cell* **94** (1998) 739–750.
116. K. Kuida, T.S. Zheng, S. Na, C. Kuan, D. Yang, H. Karasuyama, P. Rakic and R.A. Flavell, Decreased apoptosis in the brain and premature lethality in CPP32-deficient mice, *Nature* **384** (1996) 368–372.
117. Y. Liang, C. Yan and N.F. Schor, Apoptosis in the absence of caspase 3, *Oncogene* **20** (2001) 6570–6578.
118. E.E. Varfolomeev, M. Schuchmann, V. Luria, N. Chiannilkulchai, J.S. Beckmann, I.L. Mett, D. Rebrikov, V.M. Brodianski, O.C. Kemper, O. Kollet, T. Lapidot, D. Soffer, T. Sobe, K.B. Avraham, T. Goncharov, H. Holtmann, P. Lonai and D. Wallach, Targeted disruption of the mouse Caspase 8 gene ablates cell death induction by the TNF receptors, Fas/Apo1, and DR3 and is lethal prenatally, *Immunity* **9** (1998) 267–276.

119. R. Hakem, A. Hakem, G.S. Duncan, J.T. Henderson, M. Woo, M.S. Soengas, A. Elia, J.L. de la Pompa, D. Kagi, W. Khoo, J. Potter, R. Yoshida, S.A. Kaufman, S.W. Lowe, J.M. Penninger and T.W. Mak, Differential requirement for caspase 9 in apoptotic pathways in vivo, *Cell* **94** (1998) 339–352.
120. L.C. Cantley, The phosphoinositide 3-kinase pathway, *Science* **296** (2002) 1655–1657.
121. B. Vanhaesebroeck and M.D. Waterfield, Signaling by distinct classes of phosphoinositide 3-kinases, *Exp. Cell Res.* **253** (1999) 239–254.
122. B. Vanhaesebroeck, S.J. Leevers, G. Panayotou and M.D. Waterfield, Phosphoinositide 3-kinases: a conserved family of signal transducers, *Trends Biochem. Sci.* **22** (1997) 267–272.
123. M.A. Lawlor and D.R. Alessi, PKB/Akt: a key mediator of cell proliferation, survival and insulin responses? *J. Cell Sci.* **114** (2001) 2903–2910.
124. J.A. Diehl, M. Cheng, M.F. Roussel and C.J. Sherr, Glycogen synthase kinase-3beta regulates cyclin D1 proteolysis and subcellular localization, *Genes Dev.* **12** (1998) 3499–3511.
125. M. Schmidt, S. Fernandez de Mattos, A. van der Horst, R. Klompmaker, G.J. Kops, E.W. Lam, B.M. Burgering and R.H. Medema, Cell cycle inhibition by FoxO forkhead transcription factors involves downregulation of cyclin D, *Mol. Cell Biol.* **22** (2002) 7842–7852.
126. S. Fernandez de Mattos, A. Essafi, I. Soeiro, A.M. Pietersen, K.U. Birkenkamp, C.S. Edwards, A. Martino, B.H. Nelson, J.M. Francis, M.C. Jones, J.J. Brosens, P.J. Coffer and E.W. Lam, FoxO3a and BCR-ABL regulate cyclin D2 transcription through a STAT5/BCL6-dependent mechanism, *Mol. Cell Biol.* **24** (2004) 10058–10071.
127. P.F. Dijkers, R.H. Medema, C. Pals, L. Banerji, N.S. Thomas, E.W.-F. Lam, B.M. Burgering, J.A. Raaijmakers, J.W. Lammers, L. Koenderman and P.J. Coffer, Forkhead transcription factor FKHR-L1 modulates cytokine-dependent transcriptional regulation of p27(KIP1), *Mol. Cell Biol.* **20** (2000) 9138–9148.
128. J. Liang, J. Zubovitz, T. Petrocelli, R. Kotchetkov, M.K. Connor, K. Han, J.H. Lee, S. Ciarallo, C. Catzavelos, R. Beniston, E. Franssen and J.M. Slingerland, PKB/Akt phosphorylates p27, impairs nuclear import of p27 and opposes p27-mediated G1 arrest, *Nat. Med.* **8** (2002) 1153–1160.
129. B.P. Zhou, Y. Liao, W. Xia, B. Spohn, M.H. Lee and M.C. Hung, Cytoplasmic localization of p21Cip1/WAF1 by Akt-induced phosphorylation in HER-2/neu-overexpressing cells, *Nat. Cell Biol.* **3** (2001) 245–252.
130. S.R. Datta, H. Dudek, X. Tao, S. Masters, H. Fu, Y. Gotoh and M.E. Greenberg, Akt phosphorylation of BAD couples survival signals to the cell-intrinsic death machinery, *Cell* **91** (1997) 231–241.

131. L. del Peso, M. Gonzalez-Garcia, C. Page, R. Herrera and G. Nunez, Interleukin-3-induced phosphorylation of BAD through the protein kinase Akt, *Science* **278** (1997) 687–689.
132. A. Essafi, S. Fernandez de Mattos, Y.A. Hassen, I. Soeiro, G.J. Mufti, N.S. Thomas, R.H. Medema and E. W. Lam, Direct transcriptional regulation of Bim by FoxO3a mediates STI571-induced apoptosis in Bcr-Abl-expressing cells, *Oncogene* **24** (2005) 2317–2329.
133. A. Brunet, A. Bonni, M.J. Zigmond, M.Z. Lin, P. Juo, L.S. Hu, M.J. Anderson, K.C. Arden, J. Blenis and M.E. Greenberg, Akt promotes cell survival by phosphorylating and inhibiting a Forkhead transcription factor, *Cell* **96** (1999) 857–868.
134. S. Pugazhenthi, A. Nesterova, C. Sable, K.A. Heidenreich, L.M. Boxer, L.E. Heasley and J.E. Reusch, Akt/protein kinase B up-regulates Bcl-2 expression through cAMP-response element-binding protein, *J. Biol. Chem.* **275** (2000) 10761–10766.
135. J.M. Wang, J.R. Chao, W. Chen, M.L. Kuo, J.J. Yen and H.F. Yang-Yen, The antiapoptotic gene mcl-1 is up-regulated by the phosphatidylinositol 3-kinase/Akt signaling pathway through a transcription factor complex containing CREB, *Mol. Cell Biol.* **19** (1999) 6195–6206.

3 Phosphoinositide 3-kinase signalling pathways

Jan Domin

1. Introduction

Eukaryotic cell membranes are composed of a phospholipid bilayer containing an assortment of protein and carbohydrate complexes. Together they provide not only a functional barrier between cells but they also compartmentalise intracellular space into organelles. In this way, discrete microenvironments are created that contain specific transport and signalling systems. The potential for phospholipids as mediators of cell function was initially indicated in the early 1950s when phospholipid turnover was originally reported in the exocrine pancreas.[1] However, the significance of these observations were not fully realised until several years later. Although these early studies proposed an agonist-stimulated cycle of phosphatidylinositol (PtdIns) breakdown and resynthesis, this 'PtdIns cycle' was intimately associated with exocytosis.[2] By the late 1960s a calcium ion-dependent, phosphoinositide-specific, phospholipase C (PLC) activity was proposed to play a regulatory role in the PtdIns cycle in the avian salt gland and acetylcholine-stimulated brain inositol lipid hydrolysis.[3,4] The range of agonists that stimulated PLC activity began to expand rapidly[5] and the intimate relationship between intracellular Ca^{2+} levels and the PtdIns cycle was realised shortly thereafter.[6] Although PtdIns(4)P and PtdIns(4,5)P_2 were originally identified in brain extracts, they were found to have a widespread tissue distribution. The kinases responsible for their biosynthesis were found at the cytoplasmic face of the plasma membrane.[7] Stimuli that depleted PtdIns stimulated a Ca^{2+} increase and a decrease in PtdIns(4,5)P_2 levels.[8] It became clear that receptor stimulation caused a rapid, PLC-mediated hydrolysis of PtdIns(4,5)P_2 rather than PtdIns and this correlated with a rise in intracellular Ca^{2+}. The model was completed by the identification of an agonist-dependent production of Ins(3,4,5)P_3 from PtdIns(4,5)P_2[9,10] and that Ins(1,4,5)P_3 is a Ca^{2+}-mobilising second messenger[11] (Fig. 1).

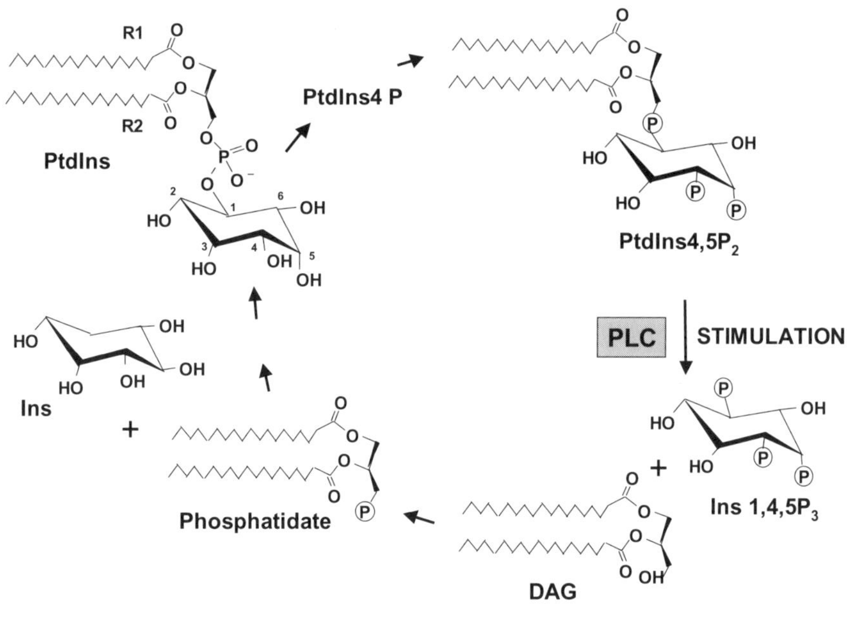

Fig. 1 A simplified phosphatidylinositol cycle. Phosphatidylinositol (PtdIns) is sequentially phosphorylated by PtdIns 4-kinase and a type I PtdInsP kinase to produce PtdIns(4,5)P_2. Agonist stimulation of receptors in the plasma membrane activate phospholipase C (PLC) which hydrolyses the inositol headgroup to generate Ins(1,4,5)P_3 and diacylglycerol (DAG) that increase intracellular calcium ions and protein kinase C, respectively. DAG is phosphorylated by a DAG kinase and the resulting phosphatidate is converted back into PtdIns by the sequential action of two enzymes. CDP-DAG synthase (CTP:phosphatidate cytidylyltransferase) catalyses the conversion of phosphatidate and CTP to form CDP-DAG. PtdIns synthase (CDP-diacylglycerol:*myo*-inositol 3-phosphatidyltransferase) then catalyses the conversion of CDG-DAG and inositol back into PtdIns.

1.1. Phosphoinositide synthesis

PtdIns is a minor component of eukaryotic cell membranes constituting approximately 8% of all phospholipids.[12] It consists of phosphatidic acid attached to an inositol ring via a hydroxyl group numbered 1′. Although this *myo*-inositol head group contains five free hydroxyl groups, due to stearic constraints only three of them (3′, 4′ and 5′) are phosphorylated *in vivo*. These products are referred to as phosphoinositides and these bioactive products are produced by a series of phosphoinositide kinases (Fig. 2). *In vitro*, phosphoinositide 3-kinase (PI3K) activity phosphorylates the 3′ hydroxyl group on the inositol headgroup of phosphatidylinositol (PtdIns), PtdIns(4)P and PtdIns(4,5)P_2 to produce PtdIns(3)P, PtdIns(3,4)P_2

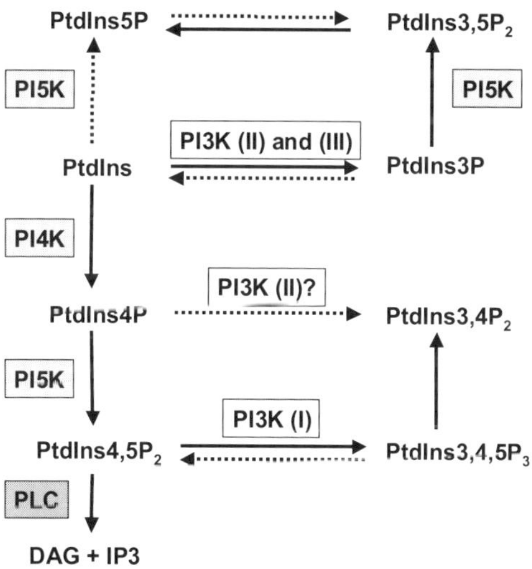

Fig. 2 Overview of 3-phosphoinositide synthesis. PtdIns is phosphorylated by a PtdIns 4-kinase to produce PtdIns(4)P which in turn is phosphorylated in the 5′ position by a type I PtdInsP kinase to generate PtdIns(4,5)P_2. PtdIns is phosphorylated by class II and class III PI3K enzymes to generate PtdIns(3)P. Hydrolysis of PtdIns(3)P regenerates PtdIns or phosphorylation by a 5-phosphoinositide kinase produces PtdIns(3,5)P_2. This product can be hydrolysed to give PtdIns(5)P which could also be produced independently by a PtdIns 5-kinase. PtdIns(4)P can act as a substrate for class II PI3K enzymes *in vitro* but it is yet unclear if this occurs*in vivo*. PtdIns(3,4,5)P3 is generated by class I PI3K enzymes in response to agonist stimulation. Once produced, PtdIns(3,4,5)P3 is hydrolysed by the action of either a 3-phosphatase or 5-phosphatases to give PtdIns(4,5)P_2 or PtdIns(3,4)P_2, respectively.

and $PtdIns(3,4,5)P_3$. These phosphoinositides then play a direct role in signal transduction and to date, no phospholipase has been described that utilises 3-phosphoinositides as substrate. In quiescent mammalian cells, the levels of $PtdIns(3,4)P_2$, $PtdIns(3,5)P_2$ and $PtdIns(3,4,5)P_3$ are low but are elevated dramatically following acute stimulation with cytokines, growth factors or stress. In contrast, both PtdIns(3)P and PtdIns(4)P are high in resting cells and their whole cell content changes little following stimulation. Whilst PtdIns(3)P and PtdIns(3,4,5)P3 are thought to be produced directly by PI3K activity, the action of 5′ phosphatases may provide most if not all $PtdIns(3,4)P_2$.[13] Similarly, 3′ phosphoinositide phosphatases convert $PtdIns(3,4,5)P_3$ into $PtdIns(4,5)P_2$. Finally, the most recently identified 3-phosphoinositide $PtdIns(3,5)P_2$ is produced by the phosphorylation of PtdIns(3)P by a PtdIns 5-kinase.[14] To learn more about PtdIns 4-kinase and PtdIns 5-kinase enzymes the reader is referred to the following reviews.[15–17] In addition to its phosphorylation by PtdIns 3-kinase activity, $PtdIns(4,5)P_2$ is also hydrolysed by members of the PLC enzyme family which generate inositol $(1,4,5)P_3$ and diacylglycerol. These metabolites result in the liberation of intracellular free Ca^{2+} and activation of protein kinase C, respectively.[18,19] Currently, the mechanism that determines whether a molecule of $PtdIns(4,5)P_2$ should serve as the substrate for either PI3K or PLC enzyme activity remains unknown, or indeed what proportion of $PtdIns(4,5)P_2$ is produced by 3′-phosphatase action.

1.2. Analysis of PI3K activity

In the early 1980s a novel kinase activity that utilised PtdIns as substrate was found to associate specifically with oncogene products such as polyoma middle T-antigen, v-src[20] and v-ros,[21] in addition with growth factor receptors such as those for platelet-derived growth factor (PDGF) and epidermal-derived growth factor (EGF).[22,23] Critically, the association of this kinase correlated with the transforming potential of each oncogene studied.[24–26] Two biochemically distinct PtdIns kinases were described in fibroblasts termed types I and II, and a type III enzyme from bovine brain.[27] Only the type I PtdIns kinases are specifically associated with activated tyrosine kinases. Type I PtdIns kinase specifically phosphorylated the 3′ position of the inositol ring to generate PtdIns(3)P.[28] In contrast, the type II PtdIns kinase in fibroblasts phosphorylated the 4′ position to produce PtdIns(4)P that was previously thought to be the only phosphoinositide that existed in cells.

1.3. Inhibitors of PI3K activity

Wortmannin and LY294002 have been extensively used as small molecule inhibitors of PI3K activity.[29,30] Together these compounds revolutionised our understanding of the biological role of 3-phoshoinositide production *in vivo*. Both compounds are cell permeable and when used at the correct concentrations (low nM for wortmannin; low μM for LY294002), each acts as a relatively specific inhibitor. However, as with all inhibitors, caution is always required when interpreting the resultant data since wortmannin also inhibits the activity of cytosolic phospholipase A2 at low concentrations.[31] Wortmannin binds covalently to the catalytic domain whilst LY294002 acts as a competitive and therefore reversible antagonist of ATP.[32] Inhibition of biological activity using wortmannin and LY294002 independently has been widely accepted as sufficient evidence to support the involvement of PI3K activity. When used at high concentrations both inhibitors induce apoptosis.[33]

Although extensively used for research purposes, the low solubility and instability of wortmannin and LY294002 in aqueous solution has precluded therapeutic application. However, the expansion of human mammary and pancreatic carcinoma cell xenografts has been reduced in SCID mice by oral administration of wortmannin.[34,35] Daily administration of LY294002 to mice suppressed growth and induced apoptosis of human colon and ovarian cancer cell xenografts.[36,37] To overcome issues relating to drug stability, several analogues of wortmannin have been developed, including PWT-458, a pegylated 17-hydroxywortmannin, which upon cleavage of its polyethyleneglycol releases 17-hydoxyywortmannin *in vivo*.[38] This compound abolished Akt phosphorylation in glioma, non-small cell lung carcinoma and renal cell carcinoma xenograft tumours in nude mice. Over the years, additional inhibitors of PI3K activity have also been described but none have been widely adopted.[39] A new generation of PI3K isoform selective inhibitors are under development. It is hoped that they will offer new therapeutic potential in diseases such as cancer that are characterised by cell hyperplasia but minimise the apoptotic cell death of normal tissue.

1.4. Characterisation

1.4.1. *Biochemical approaches*

The search to isolate the PI3K enzymes began with large-scale protein purifications of the enzyme from tissue including bovine brain. This led to the isolation of a heterodimeric complex consisting of an 85 and a 110 kDa subunit, and the sequence of each subunit was revealed by cDNA cloning.[40] Two isoforms of the p85 protein termed p85α and p85β were reported simultaneously by three groups using independent strategies.[41–43] The following year, the sequence of the catalytic subunit p110α was published.[44] However, this approach of large-scale tissue purification was to be used only again to isolate the class IB PI3K heterodimer p101/p110γ.[45]

1.4.2. *Molecular cloning*

Improvements in molecular cloning strategies and automated cDNA sequencing then led to a rapid increase in the number of PI3K catalytic subunits described confirming that these isoforms constitute a large gene family. This approach was made possible by the characterisation of a gene product in yeast whose mutation results in defects in vacuolar protein sorting (vps) and endocytosis.[46,47] The product of vps34 gene termed vps34p was shown to be a PI3K and it allowed the identification of a conserved catalytic domain and the adjacent PI kinase (PIK) homology domain. The use of degenerate oligonucleotides allowed related cDNA sequences to be amplified from reverse transcribed human RNA. These products were then radiolabelled and used as probes to screen mammalian cDNA libraries. This led to the isolation of p110β (beta), p110γ (gamma) and p110δ (delta).[48,49] Around this time, the potential of *Drosophila* as a model organism amenable to genetic modification allowing the biological significance of signalling molecules such as the PI3K enzymes began to be realized. Efforts to clone *Drosophila* PI3K enzymes identified a cDNA that encode a protein sequence that differed from both the p110 and the vps34p catalytic domain. This cDNA termed Drosophila 68_D encoded a unique PI3K enzyme.[50] Cloning from mammalian cDNA libraries identified the human isoforms PI3K-C2α, HsC2-PI3K and PI3K-C2γ[51–54] together with mouse isozymes of PI3K-C2α termed p170,[55] cpk[56] and the rat orthologue of PI3K-C2γ[57] shortly thereafter.

1.5. Nomenclature of PI3K enzymes

Cloning of mammalian p110α and yeast vps34 together with the descriptions of additional isoforms confirmed a large PI3K gene family. However, their identification by a number of independent groups began to produce names that bore no relationship to either the structure or function of the enzyme they described. It was evident that an ordered classification was required. Initial schemes included enzymes that are related to PI3Ks such as mammalian target of rapamycin mTOR[58] however, these were refined to produce one that is still in use today in which the eight mammalian PI3K isoforms are classified into three classes termed I, II and III.

2. Class IA PI3K Enzymes

From the initial protein purifications and biochemical analysis, it was clear that the catalytic activity in the p85/p110 heterodimer was contained in the 110 kDa subunit. The p85 subunit has no enzyme activity acting instead as an adaptor protein to recruit the catalytic subunit to the plasma membrane following cell stimulation.

2.1. p85 alpha

Sequence analysis of the 85 kDa adaptor subunit revealed two isoforms termed p85α and p85β.[41] p85 contains two phosphotyrosine-binding Src homology 2 (SH2) domains, two proline rich motifs, a region of homology to the breakpoint cluster region (BCR) and an N-terminal SH3 domain that binds proline-rich motifs (Fig. 3). The SH2 domains allow the interaction between p85 and phosphotyrosine residues such as those found on receptors for polypeptide growth factors including platelet-derived growth factor (PDGF) and epidermal growth factor (EGF). Biophysical analysis demonstrated that the affinity of SH2 domain binding is governed not only by the phosphotyrosine residue itself but also by the amino acid residues that

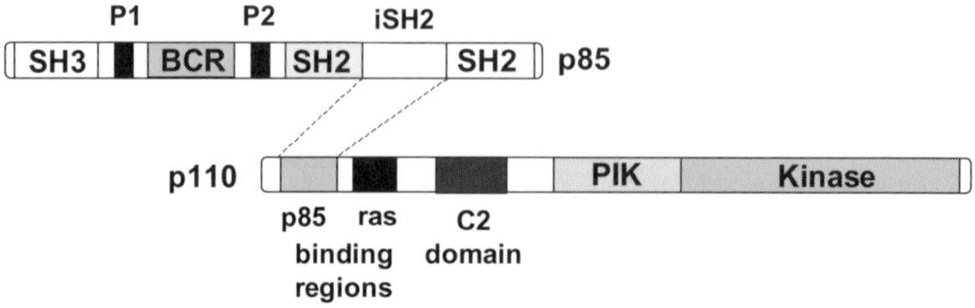

Fig. 3 Schematic representation of class IA PI3K heterodimers. The p85 adaptor consists of an N-terminal src homology 3 (SH3) domain that binds polyproline-rich motifs. p85 contains two such polyproline-rich motif termed P1 and P2 that flank a region that has homology with the breakpoint cluster region (BCR) of BCR-Abl. Tandem src homology 2 (SH2) domains allow binding to phosphotyrosine containing sequences and recruitment to receptor signalling complexes. Between the SH2 domains is an inter-SH2 sequence (iSH2) that is responsible for binding the p110 catalytic subunit. The N-terminus of p110 contains the motif that binds p85 and this lies adjacent to a sequence that binds the small GTPase ras. Within p110 is a phospholipid binding C2 domain followed by a PtdIns kinase (PIK) domain and the catalytic (kinase) domain at the C-terminus.

lie immediately C-terminal to it. For the SH2 domains of p85, a methionine residue at position +4 is preferred.[59] This interaction facilitates the recruitment of both the p85/p110 PI3K heterodimer to activated receptors. Between the two SH2 domains lies a helical region termed the inter-SH2 (iSH2) sequence that binds p110α.[60] This region contains residue serine 608 which is phosphorylated by p110 protein kinase activity.[61] At the N-terminus lies a polyproline-rich motif binding SH3 domain. Many SH3 domain-containing proteins are involved in reorganisation of the actin cytoskeleton.[62] Between the SH3 domain and the N-terminal SH2 domain is a motif with sequence similarity to the breakpoint cluster region (BCR) flanked each side by proline-rich motifs (P1 and P2). The BCR sequence was identified in patients with chronic myelogenous leukaemia in whom a chromosomal rearrangement forms the Philadelphia chromosome.[63] This creates a molecular rearrangement between the c-ABL proto-oncogene on chromosome 9 and the BCR (breakpoint cluster region) gene on chromosome 22. The BCR/ABL fusion gene codes for a novel 210 kDa protein which has higher and aberrant tyrosine kinase activity than c-ABL and results in cell transformation. Small Rho GTPase activating proteins (GAPs) also contain BCR motifs and the BCR domain of p85 has been shown to bind the small GTPases Rac and Cdc42[64,65] and regulate reorganisation of the actin cytoskeleton.[66] Each proline-rich region flanking the BCR domain could provide an intramolecular binding site for the N-terminal SH3 domain, and in this way alter conformation or serve as a binding site for regulatory proteins.

2.2. Additional class IA adaptor proteins

In total, eight isoforms of class IA adaptor have been described derived from three distinct genes p85α, p85β and p55γ. In addition to p85α and p85β, AS53 (also known as p55α[67,68]) and p50α[69,70] are splicing variants derived from the same gene encoding p85α (*Pik3r1*). They contain both SH2 domains of p85α but lack the SH3 domain, amino-terminal proline-rich domain and BCR domain. Instead, they have unique amino terminal sequences consisting of 34 and 6 amino acids, respectively. Another isoform $p55^{PIK}$ is homologous to AS53/p55α but is encoded by a different gene.[71] p85α shows the widest tissue distribution and is expressed at the highest levels.[41] In contrast, the spliced variants, AS53 and p50α, appear to play specific roles in selected tissues and are of particular relevance to signalling downstream of the insulin receptor and may play a significant role in diabetes.[72] *In vitro*, each class IA adaptor can bind each class IA catalytic subunit. However, some receptors and their adaptors appear to display a preference for specific adaptor subunits.[73]

2.3. p110 alpha, beta and delta catalytic subunits

Following the description of p110α[44] a further two PI3K enzymes were cloned. Termed p110β[48] and p110δ[49] each class IA PI3K enzyme is encoded by a separate gene. Class IA PI3K enzymes all contain an N-terminal binding site for the helical inter-SH2 domain region of p85 between residues 20–108.[60] This lies adjacent to a binding site for the small GTPase ras.[74] The catalytic and associated PIK domains form the C-terminus. Both p110α and β isoforms have a wide tissue distribution whereas the expression of the δ isoform is largely restricted to cells of haematopoietic origin.

2.3.1. *Ras binding site*

All class I PI3K enzymes bind the small monomeric G-protein ras. In p110α, the ras binding site lies between residues 133–314,[75] a region that overlaps with the ras binding site in the class IB PI3K isoform p110γ (residues 220–311). The ras binding site in PI3K enzymes has the same conformation as the ras binding site in both raf[76] and ralGDS.[77] The spatial relationship between the ras binding domain and the remainder of the enzyme suggests two alterative mechanisms by which ras binding can alter catalytic activity. In the first, ras stabilises the p110 catalytic subunit at the plasma membrane. In the second, ras binding causes a conformational change propagated through the ras binding domain to the catalytic domain to alter substrate or co-factor binding.

2.3.2. *C2 domain*

Each PI3K enzyme contains an internal C2 domain.[78,79] C2 domains, also known as CalB domains are found in a large number of proteins involved in phospholipid-mediated signal transduction.[80] These include phospholipase C (PLC), cytosolic phospholipase A2 (cPLA2) and protein kinase C (PKC).[81] It is an eight-stranded anti-parallel β-sandwich consisting of two four-stranded β-sheets. C2 domains are divided into two types; those present in enzymes such as cPLA2 and PLCδ1 whose binding to phospholipids is dependent on Ca^{2+} (type I), and those such as the C2 domain in PKCδ whose interaction with phospholipids occurs independent of Ca^{2+} (type II). The C2 domain present in the class IB PI3K p110γ is similar to the Ca^{2+}-insensitive type II C2 domain of phospholipase Cδ1.[82] Binding of C2 domains to Ca^{2+} is mediated by three loops termed CRB1, CRB2 and CRB3. The isolated C2 domain of p110γ binds multilamellar phospholipids vesicles in a similar manner to the full length enzyme in the absence of an adapter subunit.[79]

2.3.3. *Catalytic domain*

Based on the structure of p110γ, the catalytic domain consists of an N-terminal lobe and a larger C-terminal lobe.[79] The N-terminal lobe and the first part of the C-terminal lobe have a fold similar to that found in protein kinases. Both contribute to the ATP-binding site and represent the region of highest sequence similarity across all PI3K enzymes. The C-terminal lobe is involved in binding 3-phosphoinositide substrates.

2.4. Substrate specificity of class IA PI3K enzymes

In vitro class IA PI3K enzymes can all utilise PtdIns, PtdIns(4)P and PtdIns(4,5)P_2 as substrates and requires Mg^{2+} as a divalent cation. However, *in vivo*, PtdIns(3,4,5)P_3 appears to be their principle phosphoinositide product.[83] The class IA catalytic subunits also display limited serine/threonine protein kinase activity directed towards the p85 adaptor and the insulin receptor substrate-1 (IRS-1).[61] It is thought that this provides a negative feedback mechanism rather than a downstream signalling mechanism. Activation of p110δ results in autophosphorylation of this catalytic domain which serves to inhibit kinase activity.[84] The degree of sequence similarity within the kinase domain of PI3K enzymes suggests that the protein kinase activity exhibited by the class I enzymes may extend to other isoforms although potential substrates have not yet been identified.

2.5. Regulation of class IA PI3K enzyme activity

Class IA adaptor subunits are recruited to those proteins phosphorylated on tyrosine residues such as activated growth factor receptors or scaffold proteins. Translocation to the plasma membrane allows both conformational change in the PI3K heterodimer and access to phosphoinositide substrate (Fig. 4). The mechanism by which class IA PI3K heterodimers associate with activated growth factor receptors has been documented extensively and in greatest detail for the PDGF and EGF receptors.[85,86] The SH2 domain of the class IA adaptor show selectivity for the YxxM domains at tyrosine 740 and tyrosine 751 of the PDGF receptor.[87,88] Binding of the SH2 domains to their phosphopeptide substrates results in an allosteric change in the p85 adaptor that in turn is transmitted to the catalytic subunit and alters enzyme activity[89] and may induce oligomerisation of the PI3K enzyme.[90] Occupancy of both SH2 domains appears to be required for maximal activation of p110.[91] Although EGF stimulates translocation of class IA PI3K heterodimers to the activated receptor complex, the p85 subunit is thought not to bind the EGF receptor directly since it does not possess canonical p85 SH2 domain binding sites. Instead, p85 binds the docking protein Gab1[92]

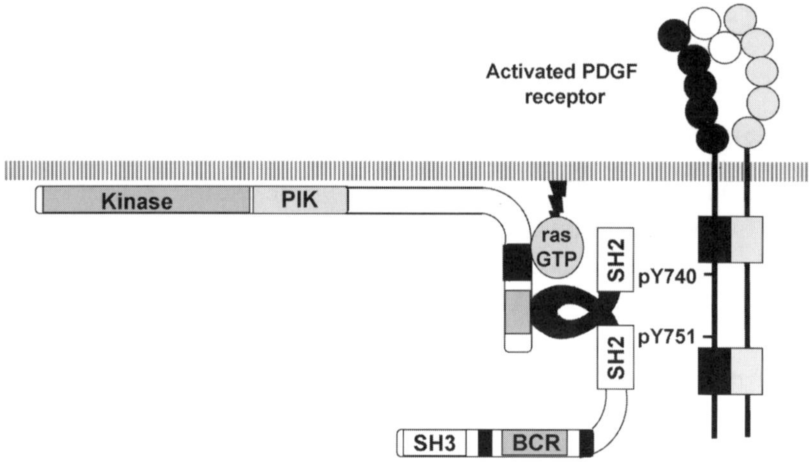

Fig. 4 Binding of the class IA PI3K heterodimer to activated platelet-derived growth factor (PDGF) receptors. Activation of PDGF receptors results in receptor dimerisation and phosphorylation of a number of tyrosine residues that include Y740 and Y751. These form binding sites for the C- and N-terminal SH2 domains of p85, respectively. Translocation of the p85/p110 heterodimer to the plasma membrane allows the p110 catalytic subunit access to its PtdIns(4,5)P_2 substrate and allows potential for conformational change that increases specific enzyme activity. At the membrane, p110 can also bind activated ras to further stimulate p110.

or if expressed, the ErbB3 receptor which has 7 YxxM motifs and forms a heterodimer with the activated EGF receptor.[93] Recruitment of class IA PI3K enzymes to a docking protein also occurs with the insulin receptor where p85 is recruited to the insulin receptor directly and both the insulin receptor substrate (IRS) and Cbl.[94,95] Binding of p85 to p110 attenuates its catalytic activity *in vitro*.[96]

Class IA PI3K heterodimers also translocate to receptors for immunoglobulin, T-cell receptors, B-cell receptors, co-stimulatory molecules such as CD28 and CD19 and integrins involved in cell adhesion and cell-cell binding. This association is facilitated by non-receptor tyrosine kinases such as the src family kinases Src, Fyn and Lyn or Janus kinases (JAK) including Tyk2, and Jak 1, Jak 2 and Jak 3.[97–99] The mechanism by which stimulatory intracellular second messengers such as the class IA PI3K enzymes were recruited to many immunoregulatory receptors remained unclear until the identification of immunoreceptor tyrosine-based activation motifs (ITAM).[100,101] This motif exists in one or more copies in each of the receptor-associated signal-transducing molecules and it consists of two tandem copies of the consensus sequence Tyr-X-X-Leu/Ile spaced by six to eight amino acids. Receptor engagement is followed by a rapid and transient phosphorylation of tyrosine residues within their ITAMs.[102]

Adhesion of cells to extracellular matrix proteins such as fibronectin results in the engagement of integrins and stimulation of 3-phosphoinositide production.[103] Integrins are heterodimers composed of one α and one β chain. There are 28 α subunits and 18 β subunits that combine in various combinations to produce 25 receptors that regulate cell adhesion, cell shape and cell motility.[104] Activation of focal adhesion complexes involves non-receptor tyrosine kinases focal adhesion kinases (FAK) and p130Cas. Integrin activation results in tyrosine phosphorylation of several protein substrates and the recruitment of class IA PI3K enzymes. Integrins are in direct contact with focal adhesion complexes which in turn regulate reorganisation of the actin cytoskeleton. Integrins also play an important role in apoptosis after the loss of cell anchorage or anoikis.[105] This process plays an important role in the life cycle of adherent cells. Furthermore, loss of anchorage dependency is believed to be a critical step in metastatic transformation.

The specific activity of p110β can also be directly increased *in vitro* by the G$\beta\gamma$ subunits of heterotrimeric G-proteins.[106,107] Following activation of G-protein coupled receptors (GPCRs) for chemoattractants, regulatory peptides or lipid mediators such as lysophosphatidic acid, they associate with the GDP-bound α-subunit of Gα/G$\beta\gamma$ heterotrimeric G-proteins. The exchange of GDP for GTP dissociates the heterotrimer into Gα and G$\beta\gamma$ subunits. Each of these is able to exert a regulatory role.

In vivo, ligands that stimulate G-protein coupled receptors have been shown to activate class IA PI3K activity, however the possibility of their transactivation of receptors for polypeptide growth factors such as that for epidermal growth factor (EGF) and platelet-derived growth factor (PDGF) is a complicating factor.[108,109]

Class I PI3K enzymes can also act as protein kinases. Amongst the limited number of substrates identified to date is the p85 adaptor.[61] Following agonist stimulation, p85α was phosphorylated in a p110α-dependent manner on serine 608 which lies in the inter-SH2 domain. Phosphorylation of p85 on this serine resulted in an 80% decrease in PI3K activity, which was subsequently reversed with protein phosphatase 2A treatment. The p110β, p110δ and p110γ isoform undergo autophosphorylation at a single c-terminal serine residue.[84,110] *In vitro* autophosphorylation of p110δ abolished lipid kinase activity. Phosphospecific antisera demonstrate that whilst phosphorylation on serine 1039 is low in quiescent cultures, it increases dramatically upon stimulation and its translocation to receptor complexes such as CD28 in T-cells.[84]

3. Class IB PI3K Enzymes

3.1. p110 gamma

p110γ represents the only class IB PI3K isoform in mammalian cells. Analogous to class IA enzymes, p110γ also exists as a heterodimer but not in complex with a class IA adaptor. Instead, p110γ binds a 101 kDa protein termed p101.[45] The X-ray crystal structure of p110γ was solved to 2.2A resolution and this has allowed insight into regulatory mechanisms of kinase activity.[79,111] Residues 1–143 of p110γ contain the binding site for the adaptor protein p101.[112] This location is equivalent to the p85 binding site of the class IA PI3K enzymes. p110γ also contains a ras-binding domain, a C2 domain, a helical domain and a catalytic domain. The helical domain contains a fold similar to HEAT repeats contained in structures involved in protein-protein interactions.[113] Stimulation of p110γ results in activation of phosphoinositide and protein kinase activity and results in the autophosphorylation of p110γ on serine 1101.[114] In contrast to the class IA PI3K isoforms, autophosphorylation of p110γ did not decrease lipid kinase activity and autophosphorylation is significantly enhanced by G$\beta\gamma$ in a time- and concentration-dependent manner. G$\beta\gamma$ subunits recruit p110γ from the cytosol to the membrane via the p101 subunit.[115]

3.2. p101 and p84 class IB adaptor proteins

p101 has no sequence similarity to other class I PI3K adaptors. p101 confers the sensitivity of the p101/p110γ complex to G-protein stimulation.[45,116] Recombinant Gs, Gi, Gq, or Go alpha subunits activated with GTP were unable to activate p110γ whilst Gβ (1–4) but not Gβ5 in complex with gamma2-stimulated lipid kinase activity about 26-fold. The heterodimers $G\beta1\gamma2$, $\beta1\gamma10$, $\beta1\gamma12$ and $\beta1\gamma13$ dimers but not $\beta1\gamma11$ also activated p110γ.[117] A second adaptor for p110γ termed p84 was recently described.[118] p84 is broadly expressed in cells of the murine immune system. Like p101, binding of p84 to p110γ increases the ability of $G\beta\gamma$ to stimulate lipid kinase activity both *in vitro* and *in vivo*. However, the p84/p110γ heterodimer is approximately four-fold less sensitive to $G\beta\gamma$ than the p101/p110γ heterodimer. In this way, p110γ appears to have two regulatory subunits, however their relationship is unclear.

3.3. Regulation of class IB PI3K activity

The p101/p110γ heterodimeric complex is activated by the $G\beta\gamma$ subunit of heterotrimeric G-proteins formed after stimulation of G-protein coupled receptors such as those for peptides and chemokines. It has also been demonstrated that the $G\alpha$ subunit may also play a regulatory role.[116,119] Characterisation of the G protein-linked receptor-mediated activation of PI3K signalling implicates $G\alpha q$, $G\alpha i$ and beta gamma dimers, but not $G\alpha s$ or $G\alpha 12$.[119] A series of deletion mutants have demonstrated that large areas of p101 including both the N- and C-terminal portions interact with the N-terminus of p110γ.[113] Furthermore, the binding of p110γ to the N-terminal region of p101 was indispensable for activation with $G\beta\gamma$.

A mutant form of p110γ that no longer phosphorylated phosphoinositides but retained protein kinase activity has been described.[120] $G\beta\gamma$ subunits stimulate autophosphorylation of p110γ which can also occur in the absence of p101.[121] However, p101 alters the phosphoinositide substrate preference of p110γ following $G\beta\gamma$ stimulation. Using PtdIns(4,5)P_2 as substrate, $G\beta\gamma$ potently activates the p101/p110γ heterodimer but negligibly stimulates the p110γ monomer. In addition to its autophosphorylation following stimulation, p110γ can phosphorylate both p101 and the protein kinase MEK-1.[120,122] This effect is also differentially regulated by p101.

4. Class II PI3K Enzymes

There are three class II PI3K enzymes in mammalian cells termed PI3K-C2α, PI3K-C2β and PI3K-C2γ (Fig. 5). Each is encoded by a separate gene. The class II PI3K enzymes are characterised by their high molecular

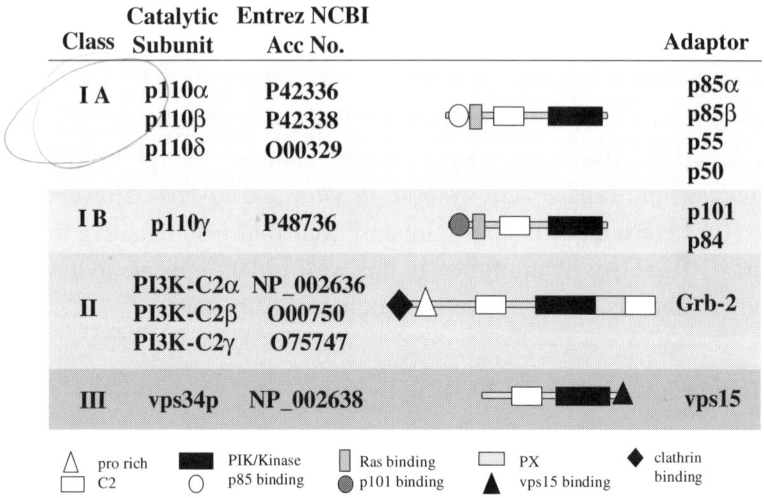

Fig. 5 The PI3K enzyme family. Eight mammalian PI3K isoforms have been organised into three classes termed I, II and III based on sequence and phosphoinositide substrate specificity. The class I isozymes are further divided into those that bind the p85 adaptor (class IA) and those that bind p101 (class IB).

mass >170 kDa and are defined by two phospholipid binding motifs in their defining C-terminal region.

4.1. Structure

All class II PI3K enzymes contain the internal C2 domain, a PIK domain and the catalytic domain described above for the class I PI3K enzymes. These regions provide areas of greatest sequence similarity with the class I and class III PI3K enzymes. However, in addition, class II PI3K enzymes contain a number of additional domains that confer distinctive mechanisms of enzyme regulation. Because of their significantly larger molecular mass the possibility exists that in contrast to the class I and class III PI3K enzymes intrinsic conformational structure may regulate class II PI3 enzyme activity.

4.1.1. *C-terminal C2 domain*

The first phospholipid binding domain identified in the class II PI3K enzymes was a type II C2 domain at the C-terminus.[50] This C-terminal C2 domain binds phospholipid vesicles in a Ca^{2+}-independent manner consistent with the absence of a critical Asp residue in the third CRB loop (CRB3). A PI3K-C2β mutant with a deletion of the C-terminal C2 domain increased the specific activity of this enzyme using PtdIns as substrate indicating a regulatory role.[53] Despite its ability to bind phospholipids, deletion of the C2 domain did not alter the subcellular localisation of PI3K-C2α.[123]

4.1.2. *Phox homology (PX) domain*

In neutrophils, a multicomponent NADPH oxidase system produces large quantities of reactive oxygen species (ROS) in a process termed respiratory burst. The minimal core of NADPH oxidase is composed of two membrane-bound subunits of cytochrome b(558) [gp91(phox) and p22(phox)] and three soluble factors [GTP-Rac, p47(phox) and p67(phox)]. p40(phox) is a binding partner for p67(phox) and allows PtdIns(3)P to stimulate ROS production. Binding of PtdIns(3)P to p40(phox) is mediated by its PX domain.[124–126] Extending the number of PX domains examined revealed that PX domains from different proteins showed varied specificity for phosphoinositide binding.[127] The PX domain of cpk, the murine homologue of PI3K-C2α, showed a preference for PtdIns(4,5)P_2.[128] Deletion of the PX domain of PI3K-C2α also did not alter enzyme localisation[123] consistent with the observation that translocation of p47(phox) PX domain to the plasma membrane involves 3′-phosphoinositide(s), but phosphoinositide binding of p47(phox) PX domain is not sufficient to recruit it to the plasma membrane.[129]

4.1.3. *Proline-rich motifs*

The N-termini of class II PI3K enzymes show no sequence similarity to any other proteins, and in contrast to the class I PI3K enzymes lack a ras-binding consensus sequence. In contrast, they contain a number of proline-rich motifs that have been shown to bind the SH3 domain containing adaptor protein Grb-2.[130] This interaction allows the association of PI3K-C2β to bind the epidermal growth factor receptor through specific phosphotyrosine residues.[131] In an attempt to identify regulators of class II PI3K enzyme activity, we are currently screening for additional binding partners for these proline-rich regions.

4.2. Regulation of kinase activity

Class II PI3K enzymes phosphorylate both PtdIns and PtdIns(4)P *in vitro* and only phosphorylate PtdIns(4,5)P_2 poorly under conditions of altered phosphoinositide presentation.[51] Importantly, class II PI3K enzymes are uniquely able to phosphorylate phosphoinositides using Ca^{2+} as an alternative to Mg^{2+} as the divalent cation *in vitro*.[53,131] It remains unclear if this specificity is physiologically relevant but it provides a convenient way to assess the contribution of class II PI3K enzyme activity to 3-phosphoinositide production without specific immunoprecipitation of each isoform. Uniquely, the class II PI3K isoform PI3K-C2α is refractory to the inhibitory effects of wortmannin being approximately ten-fold less sensitive than the other PI3K enzymes.[51]

Until recently, the 3-phosphoinositides produced by class II PI3K enzymes *in vivo* remained controversial. In contrast to class I PI3K enzymes, transient expression of recombinant class II PI3K enzymes did not increase production of either PtdIns(3,4)P_2 or PtdIns(3,4,5)P_3. However, several studies have indicated that class II PI3K enzymes produce PtdIns(3)P *in vivo*. In human platelets, activation of integrin alphaIIb beta3 and its subsequent binding to fibrinogen, stimulates synthesis of PtdIns(3,4)P_2. This synthesis depends upon the generation of PtdIns3P and its phosphorylation by a PtdIns(3)P 4-kinase.[132] PI3K-C2β kinase activity was stimulated in lysates from platelets aggregated in an αIIb β3-dependent manner or incubated with Ca^{2+}.[133] Highly purified liver nuclei incorporated radiolabelled phosphate into PtdIns(4)P, PtdIns(4,5)P_2 and PtdIns(3,4,5)P_3. Similarly, when nuclei are depleted of their membrane, they lack class I PI3K enzymes. When nuclei were harvested 20 hours after hepatectomy and treated in this way, generation of PtdIns(3)P could be observed together with an increase in PI3K-C2β activity.[134] Furthermore, analysis of 3-phosphoinositide production in nuclei following cell cycle progression showed that an increase in PtdIns(3)P levels correlated with an increase in the specific activity of PI3K-C2β.[135] Stimulation of renal cortical slices with hepatocyte growth factor transiently elevates PtdIns(3)P production in brush-border plasma membranes and this effect also increased PI3K-C2β activity in this fraction.[136] Most recently, PI3K-C2β has been shown to stimulate cell migration via a PtdIns(3)P-dependent pathway. In contrast, the expression of a kinase inactive form of PI3K-C2β inhibited

PtdIns(3,4,5)P_3-dependent phosphorylation of Akt and cell growth in a small cell cancer cell line leading to the conclusion that PI3K-C2β contributes to PtdIns(3,4,5)P_3 production *in vivo*.[137]

4.3. Role of class II PI3K in vesicle transport

In contrast to class I PI3K enzymes that are predominantly cytosolic in quiescent cells, class II PI3K enzymes are associated with phospholipids in plasma membrane and low density microsomes[53,123] To date, most of what we understand about the role of class II PI3K enzymes in vesicle transport was achieved using PI3K-C2α. Immunofluorescence staining of PI3K-C2α revealed that the enzyme had a punctate distribution with a perinuclear localisation.[123] To define this localisation further, PI3K-C2α immunoreactivity was enriched upon purification of clathrin-coated vesicles.[123,138] In addition, both PI3K-C2α and PI3K-C2β bind clathrin directly via clathrin-binding motif at their N-termini (Wheeler and Domin, *J Cell Physical*, in press). In this way, clathrin functions as an adaptor for class II PIK enzymes and stimulating their catalytic activity. Further, exogenous expression affects clathrin-mediated endocytosis and sorting in the trans-Golgi network.[139] We have recently described a role for PI3K-C2α and PtdIns(3)P in calcium-dependent fusion of secretory vesicles with the plasma membrane in neuroendocrine cells (Meunier *et al.*, *Mol Biol Cell*, in press). Impairment of PI3K-C2α kinase activity or sequestration of PtdIns(3)P abolished secretion, and in this way both are required for fusion competence in neurosecretion.

4.4. Role of class II PI3K enzymes as downstream targets of growth factor receptors, chemokine receptors and integrins

Isolation of activated polypeptide growth factor receptors for insulin,[139] epidermal growth factor and platelet-derived growth factor[131] have all revealed a co-immunoprecipitation of class II PI3K enzymes in a manner analogous to class IA PI3K enzymes. The association of PI3K-C2β with the EGF receptor is mediated by the Grb-2 adaptor and requires phosphorylation of EGF receptor on residues tyrosine 992, tyrosine 1068 and tyrosine 1173.[131] Stimulation of the macrophage chemotactic peptide-1 (MCP-1) receptor and αIIIβV integrin by fibronectin increases the specific activity of PI3K-C2α and PI3K-C2β, respectively.[133,140] The mechanism responsible for this effect is unclear and it has proved difficult to demonstrate a similar response following growth factor receptor stimulation.

4.5. Biological roles for class II PI3K enzymes

Studies in Drosophila demonstrated that when expressed in precursors that form the wing and thorax, the class II PI3K enzyme alters pattern formation but not growth or organ size.[141] This indication that class II PI3K enzymes might play a role in cell differentiation and development was supported by analysis of class II PI3K enzyme distribution in human tissue.[142] These data indicate that class II PI3K enzymes are expressed selectively in fully differentiated, non-proliferating cells and absent in stem cell/progenitor compartments. Using enzyme over-expression and RNAi, it was shown that PI3K-C2β plays a role in cell migration and reorganisation of the actin cytoskeleton in a PtdIns(3)P-dependent manner (Domin *et al.*, *J Cell Physiol*, in press and Maffucci *et al.*, *J Cell Biol*, in press). These findings may have important implications for the progression of acute and chronic inflammation, atherosclerosis and tumour metastasis. In contrast, attenuation of PI3K-C2α expression using RNAi results in cell death.[143]

4.6. Tissue distribution

Of the class I PI3K enzymes and adaptor subunits, p110α and p110β have a widespread tissue distribution. This contrasts with p110δ and p110γ which are predominantly expressed in haematopoietic cells. The class IA PI3K adaptor subunits also show differential expression with p50α expressed mainly in brain, muscle, liver and kidney,[67] whereas p55α is expressed mainly in brain and muscle alone.[144] PI3K-C2α and PI3K-C2β isoforms have a widespread distribution amongst mammalian tissues. In humans, both PI3K-C2α and PI3K-C2β are expressed in selected cell populations of epithelial or mesenchymal origin.[145] Both isoforms are present in columnar epithelium and transitional epithelia but absent in mucous-secreting and stratified squamous epithelia. Mesenchymal elements (smooth muscle and endothelial cells) and glomerular epithelium only express PI3K-C2α, while ganglion cells expressed PI3K-C2β. Both isoforms were detected in macrophages. In contrast, PI3K-C2γ is expressed predominantly in the liver.[57]

5. Class III PI3K Enzymes

The first class III PI3K enzyme was identified using a genetic screening programme in yeast to identify proteins involved in vacuolar protein sorting (vps).[145,146] Of the series of vps mutants identified, the gene product of vps34 termed vps34p was found to have sequence homology with the then recently identified p110α class IA catalytic subunit.[46] In yeast vps34p produces PtdIns(3)P, it has been largely assumed that this PI3K isoforms is responsible for most if not all the generation of PtdIns(3)P in mammalian cells. The steady state levels of PtdIns(3)P alter little following receptor stimulation, leading many authors to conclude that activation of this PI3K isoform is constitutive.

5.1. Vps34

Vps34p is of a similar molecular mass to the class I PI3K enzymes. Its sequence similarity with the class I PI3K isoforms decreases outside the PIK/kinase domain and C2 domain. The C-terminus contains a region involved in binding the class III PI3K adaptor vps15p and this association is dependent on vps15p kinase activity.[147]

5.2. Vps15p

In both yeast and mammalian cells, the class III PI3K exists in complex with a serine/threonine protein kinase termed vps15p in yeast[148] and p150 in humans.[149] Both yeast and human isoforms are post-translationally modified by the addition of a myrisoyl moiety at their N-terminus to target the class III PI3K enzyme complex to the membrane. The N-terminus of vps15p contains the kinase domain and the binding site for vps34p. Towards the C-terminus, vps15 contains HEAT repeats that mediate its interaction with vps34p[147] and WD40 domains that bind the small GTPase Rab5.[150]

5.3. Regulation of class III PI3K enzyme activity

Studies in yeast confirm that vps34p produces PtdIns(3)P *in vivo*. As such, the class III PI3K homologue has been accredited with the constitutive production of PtdIns(3)P in cells. The interaction of vps34p with vps15p increases lipid kinase activity.[149] Both vps34p and vps15p exist as phosphoproteins *in situ*. Expression of vps15p containing a point mutation precluding kinase activity inhibits autophosphorylation and results in defects in vacuolar sorting.[148,151] As with the class IA PI3K adaptor p85, vps15p has been shown to exist free from vps34p.[150] This raises the possibility that vps15 may have functions independent of vps34p.

Whilst yeast has provided fantastic insight, sufficient differences exist to make direct extrapolation between the activity of vps34p in yeast and mammals difficult. Nevertheless, the work in yeast has hindered progress examining the role of vps34p in mammalian cells.

5.4. Biological role of class III PI3K enzymes

Although in mammalian cells, all 3-phosphoinotides are likely to play a role in some aspects of membrane transport, most attention has focussed on PtdIns(3)P.[152] In *Saccharomyces cerevisiae*, mutants of vps34p show defects in membrane traffic.[46,146] Vps34 is best characterised as the regulator of carboxypeptidase Y pathway that delivers newly synthesised proteins such as the carboxypeptidase Y hydrolase from the transGolgi network to an endosomal compartment and later to the vacuole.[46] In addition to vps15p which recruits and activates vps34p at the Golgi/endosome multiple proteins are required to transport carboxypeptidase Y, including vac1/vps19, vps27 and vam7.[153–155] Each contains a PtdIns(3)P binding FYVE (vac1 and vps27) or PX (vam7) domain. Mutations in either domain disrupt carboxypeptidase Y transport.[156]

Vps34p also regulates autophagy,[157,158] a non-classical membrane trafficking pathway that transports proteins and organelles from the cytoplasm into the lumen of the vacuole/lysosome.[159] Autophagy is dependent on PtdIns(3)P production[160] and vps34p co-immunoprecipitates with proteins agp6/vps30 and agp14 that regulate this process.[157] Surprisingly, FYVE domain-containing proteins do not appear to be involved.[161] Instead a protein termed Etf1 was identified that binds PtdIns(3)P through a basic amino acid motif (KKPAKK).

Although yeast lack a class IA PI3K enzyme, fission yeast *Schizosaccharomyces pombe* express ptn1; an orthologue of PTEN a 3-phosphoinositide phosphatase that acts on PtdIns(3,4,5)P_3 in higher eukaryotes.[162] Deletion of ptn1 increased PtdIns(3,4,5)P_3 to levels comparable with those of mammalian cells, demonstrating that ptn1 suppresses PtdIns(3,4,5)P_3 synthesis. In *S. pombe*, PtdIns(3,4,5)P_3 is produced by Vps34p and the phosphatidylinositol-4-phosphate 5-kinase, its3p, but did not involve the phosphatidylinositol-3-phosphate 5-kinase, fab1p.[163,164] This indicates that a pathway for PtdIns(3,4,5)P_3 synthesis downstream of a class III phosphoinositide 3-kinase evolved before the appearance of the class I phosphoinositide 3-kinase enzymes. In addition to their role in vesicle transport, vps34p is also considered to play a role in receptor-mediated signalling downstream of the transforming growth factor-β (TGFβ) receptor since translocation of the Smad anchor for receptor activation (SARA) is mediated by PtdIns(3)P production.[165]

6. Expression of PI3K Enzymes

6.1. Subcellular localisation

Translocation of the class IA PI3K enzymes from the cytoplasm to activated receptors at the plasma membrane provides a paradigm by which PI3K signalling is stimulated. Association with the p85 adaptors regulates kinase activity of class IA enzymes since over-expression of the catalytic subunits alone increases intracellular levels of PtdIns(3,4,5)P_3 and stimulates the phosphorylation of downstream targets. The avian retroviral oncogene gene V-p3k, a homologue of p110α induces hemangiosarcomas in chickens.[166] The fusion of v-p3k to viral gag proteins results in the constitutive localisation of the gene product to the plasma membrane and correspondingly increases 3-phosphoinositide levels. Similarly, membrane-targeted p110α stimulates cell cycle progression and results in a transformed phenotype.[167]

In contrast, both the class II and class III PI3K enzymes appear to be constitutively associated with phospholipids membranes.[123,168] Consequently, for these isoforms translocation of each enzyme is not an issue but rather the regulation of their catalytic activity by receptor recruitment in the case of the class II PI3K enzymes[169] or phosphorylation by vps15p in the case of vps34p.

6.1.1. *Nucleus*

In recent years, increasing evidence is accumulating to demonstrate a role for 3-phosphoinositides in the nucleus.[170] Originally this was based on biochemical fractionations that were prone to contamination by other cellular compartments. Antisera raised to 3-phosphoinositides demonstrated a nuclear localisation *in situ*.[171] In this way, PtdIns(3,4)P_2 was localised to the speckle domains of nuclei.[172] These regions contain components of the transcriptional and pre-RNA processing cascade. Another approach utilised PtdIns(4,5)P_2-binding PH domains from proteins such as phospholipase $C\delta_1$.[173,174] The type I phosphatidylinositiol phosphate kinase (PIP kinase) that phosphorylates PtdIns(4)P on the 5′ phosphate and produces PtdIns(4,5)P_2 is also localised in the nucleus. It is estimated that the nuclei contain approximately 20% of the total cellular PtdIns(4,5)P_2 content. Stimulation of PC12 cells with nerve growth factor increased levels of class IA PI3K enzyme, their activity and PtdIns(3,4,5)P_3 generation in nuclei.[175,176] Regulation of class IA PI3K enzyme activity in the nucleus is regulated by a GTPase termed PIKE.[177] In contrast to regulation by ras in the cytoplasm, activation by PIKE occurs via the p85 subunit. In quiescent HepG2 cells, p110γ displays a cytoplasmic localisation, translocating to the nucleus upon serum stimulation.[178] Translocation of p110γ to the nucleus is inhibited by pertussis toxin and this effect could also be induced by over-expression of $G\beta\gamma$ alone. When expressed as a recombinant protein, a p110γ mutant that did not associate with p101 was constitutively localised to the nucleus.

The class II PI3K isoforms, PI3K-C2α and PI3K-C2β, have both been described in the nucleus. PI3K-C2α was localised to speckles by immunofluorescence staining of HeLa cells and the nuclear localisation signal characterised in the C2 domain.[179] The specific activity of PI3K-C2β increased during cell cycle progression of HL60 cells. Vps34 has been described in plant cell nuclei at sites of transcription.[180]

Unfortunately, the exact role of phosphoinositides in the nucleus currently remains poorly defined. PtdIns(4,5)P_2 influences chromatin structure by regulating the interaction between nuclear matrix and the chromatin remodelling complex BAF to control nuclear actin.[181] In addition, PtdIns(4,5)P_2 is thought to alter pre-mRNA splicing in nuclear speckles.[172] Similarly, PtdIns(3,4)P_2 was also detected in the nuclei.[182]

6.2. Tissue distribution

Whereas p110α and p110β are widely expressed, the expression of p110δ and p110γ is predominantly localised to haematopoietic cells. Of the class IA adaptor subunits, p55α is expressed mainly in the brain and muscle,[68] whereas p50α is expressed in the brain, muscle liver and kidney.[70] In humans, the class II PI3K isoforms PI3K-C2α and PI3K-C2β are primarily expressed in selected cell populations of epithelial or mesenchymal origin and mononuclear leukocytes.[142] These include columnar epithelium and transitional epithelia. Smooth muscle, endothelial cells and glomerular epithelium only express PI3K-C2α whilst ganglion cells express PI3K-C2β. Neither class II PI3K isozyme was detected in the stem cell/progenitor compartments or in B lymphocyte aggregates. To date, the tissue distribution of vps34p remains unknown.

6.3. Phylogenetic expression

At least one representative from each class of PI3K enzyme is found in metazoa. The slime mould *Dictyostelium discoideum* has three isoforms related to the class IA PI3K enzymes. A single class IA PI3K heterodimer is present in the nematode *Caenorhabditis elegans* (AGE-1/AAP-1) and fruitfly*Drosophila melanogaster* (Dp110/p60). Analysis of *C. elegans* mutants revealed that dauer formation, stress resistance and longevity are determined in part by a DAF-2 (insulin receptor-like protein), AGE-1 (PI3K) and DAF-16 (FOXO transcription factor) pathway. In *Drosophila*, each PI3K class is represented by a single enzyme. Mammalian cells appear to express at least one class IA PI3K isoform. A single class II PI3K enzyme is expressed in *Drosophila* and *C. elegans* but they are absent in yeast, *Dictyostelium* and plants. Only a class III PI3K is found in unicellular eukaryotes such as *Saccharomyces cerevisiae* and plants, reflecting their ability to produce PtdIns(3)P.

6.4. Related enzymes

Analysis of primary amino acid sequence identified significant sequence similarity across the catalytic domain between PI3K, PI4K enzymes and protein kinases, that include the yeast Target of Rapamycin (TOR) enzymes TOR1 and TOR2 together with mammalian orthologues FRAP1 and RAFT1,[183–185] DNA-dependent protein kinase (DNA-PK) and Ataxia Telangiectesia mutated (ATM).[186,187] Indeed initially these were included in several schemes of PI3K enzyme classification. However, currently there is no evidence to support their role as phosphoinositide kinases.

7. 3-Phosphoinositide Binding Domains

Despite many years of intense research effort and the identification of a PI3K gene family, the identification of the mechanism by which 3-phosphoinositides achieved their downstream effects remained elusive. Initial attempts involved the use of many varieties of assays based on binding affinity. Interestingly, despite these screening programmes, the major breakthrough came with the analysis of proteins whose activity or localisation was dependent on PI3K activity (Fig. 6).

7.1. FYVE domains

FYVE is an acronym based on four proteins whose binding to phospholipids vesicles was found to be dependent on PI3K activity. These proteins were **F**ab1p, **Y**OTB, **V**ac1p and **E**arly endosomal antigen (EEA1).[155,188] FYVE domains exhibit a binding specificity for PtdIns(3)P and have been identified in 27 proteins in the human genome.[189] This domain is approximately 60 amino acid residues long and the crystal structure from vps27p reveals a compact core of two small double-stranded β-sheets, a long N-terminal loop and a C-terminal α-helix. Zn^{2+}-chelating Cys/His residues. The FYVE domain surface contains a relatively large basic region contributed by a conserved RKHHCR motif located near and on b1, and by a conserved arginine from the b4 strand-charged patch. Mutagenesis of the RKHHCR motif results in loss of PtdIns(3)P binding.[155] PtdIns(3)P binds within a binding pocket that does not allow binding of higher 3-phosphoinositides

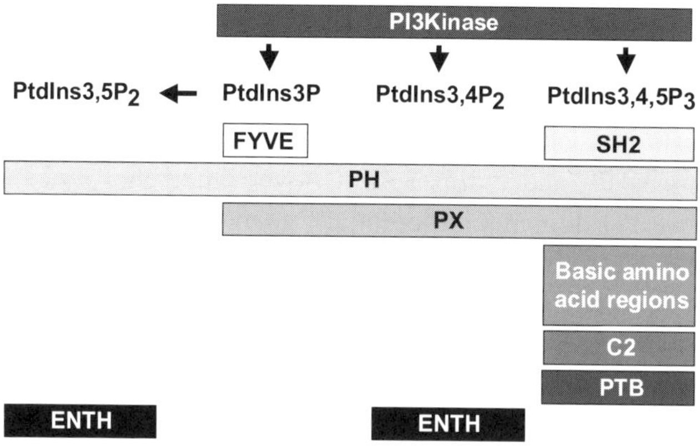

Fig. 6 Schematic of 3-phosphoinositide binding domains. Stimulation of PI3K enzyme activity produces PtdIns(3)P, PtdIns(3,4)P_2, PtdIns(3,4,5)P_3 and PtdIns(3,5)P2. Each of these 3-phosphoinositide products binds a number of protein motifs that include the **F**ab1p, **Y**OTB, **V**ac1p and **E**arly endosomal antigen (EEA1) domain (FYVE), src homology 2 domain (SH2), pleckstrin homology domain (PH), Phox homology domain (PX), C2 domain, phosphotyrosine binding domain (PTB) and the epsin N-terminal homology domain (ENTH).

or PtdIns(4)P.[190] This interaction frequently localises proteins to endosomes on which PtdIns(3)P is concentrated.[191]

7.2. PH domain

The plekstrin homology domain was originally identified using computer-based protein modelling since sequence alignment of these domains reveals a single conserved tryptophan residue.[192–194] Although originally proposed to bind small GTPases, subsequent analysis of its binding specificity revealed its ability to bind inositol phosphate headgroups.[195] PH domains are approximately 100 amino acid residues long and in contrast with FYVE domains, they are found in a far wider selection of proteins. The binding specificity of PH domains towards 3-phosphoinositides is more promiscuous than that displayed by FYVE domains and functionally, PtdIns(3,4)P_2 and PtdIns(3,4,5)P_3 mediated translocation and activation of the Akt kinase is dependent on its pleckstrin homology domain.[196–198] Despite the fact that different PH domains exhibit differential binding to 3-phosphoinositides,

PH domains do not bind PtdIns(3)P with high affinity, the binding of PH domains to PtdIns(3)P allows the C2K interacting protein (CKIP-1) to translocate to the plasma membrane and regulate the differentiation of myoblasts into muscle.[199] A subgroup of PH domains preferentially bind PtdIns(3,4)P_2 and PtdIns(3,4,5)P_3 over other phosphoinositides but *in situ*, these PH domains preferentially bind PtdIns(3,4,5)P_3. To date, a PtdIns(3,4)P_2-selective binding domain has not been identified. Its absence suggests that PtdIns(3,4)P_2 may lack a specific biological function distinct from PtdIns(3,4,5)P_3 and by inference suggest that no PI3K isoform selectively generates this 3-phosphoinositide.

PH domains are composed of two opposed β-sheets arranged at an angle of approximately 60° forming a barrel-like structure capped at one end by an α-helix. This together with each β-sheet forms the 3-phosphoinositide binding site. Only subsets of PH domains display a high affinity binding to phosphoinositides and the residues that dictate this specificity are located within their N-terminus. A recent report describes a 'split' pleckstrin homology (PH) domain derived from phospholipase C-γ1 and the TRPC3 ion channel.[200] This provides a novel mode of ligand interaction in which two PH domain fragments in distinct proteins associate to form an 'intermolecular' PH domain that binds inositol phospholipids and regulates ion channel location and function.

7.3. PX domain

This domain was identified following analysis of the phox proteins involved in the respiratory burst of neutrophils, and derives its name from the phagocyte oxidase (PhoX) complex required for killing of phagocytosed pathogens and is associated with chronic granulomatous disease in man.[201] The PH domains of p40 phox and p47 phox bind PtdIns(3)P and PtdIns(3,4)P_2 whilst the PH domains of the sorting nexin family (SNX) are generally selective for PtdIns(3)P. The production of specific phosphoinositides at intracellular membranes provides docking sites for proteins containing FYVE domains, PH domains or PX domains to phospholipid membranes. This binding translocates proteins from one intracellular compartment to another, it can alter protein conformation and bring proteins in close proximity to each other.

7.4. ENTH domain

The epsin N-terminal homology (ENTH) domain is a region of approximately 140 amino acid residues that is well conserved in eukaryotes, it is found in a number of proteins involved in vesicle transport and reorganisation of the actin cytoskeleton.[202,203] These proteins include Epsin and its homologues, adaptor protein 180 (AP180), Huntingtin-interacting protein 1 (HIP1) and the clathrin assembly lymphoid myeloid leukaemia protein (CALM).[204] ENTH domains bind PtdIns(4,5)P_2,[205,206] and the ENTH domain of HIP1 and HIP1-related binds PtdIns(3,5)P_2 and PtdIns(3,4)P_2.[207,208] Solution structure of several ENTH domains has been solved and shows a series of eight to ten α-helices, the first seven of which form the phosphoinositide-binding site in CALM, whilst the 3-phosphoinositide-binding site in epsin was mapped to an unstructured region that precedes the first α-helix.[209] The spectrum of 3-phosphoinositides that bind ENTH domain-containing proteins provides a clear link between regulation of vesicle transport and the reorganisation of actin filaments.

8. Downstream Targets of 3-Phosphoinositides

8.1. PtdIns(3)P

The PtdIns(3)P-binding FYVE domains are found in approximately 60 proteins that play a role in secretory and endocytic pathways. In addition to Fab1, YOTB, Vac1p and Early endosomal antigen (EEA1), FYVE domains are present in proteins such as the mammalian hepatocyte receptor tyrosine kinase substrate (Hrs), myotubularin related protein 3 (MTMR3) and Smad anchor for receptor activation (SARA). Fab1 and its human orthologue p235 are PtdIns(3)P-5′ kinases potentially regulated downstream of cytokine and growth factor receptors. They are involved in the formation of PtdIns(3,5)P_2[210] and multivesicular endosomes and membrane recycling from the lysosome/vacuole.[211,212] YOTB is a hypothetical gene product in *C. elegans*, related to the Fgd family of proteins in mammalian cells that are GEFs for Cdc42. YGL023 (PUF4) is a member of a protein family that preferentially binds mRNAs encoding nucleolar ribosomal RNA-processing factors. vac1 and its mammalian orthologue EEA1 is associated with endosomal compartments, vps45 an orthologue of Sec1 and vps21, a Rab GTPase related to human Rab5.[213]

Down-regulation of cell surface proteins depends on the function of the multivesicular body (MVB) sorting pathway. This pathway requires the class E vacuolar protein sorting vps proteins that form the ESCRT-I complex (vps23p, vps28p and vps37p) and vps27p (the yeast homologue of mammalian hepatocyte receptor tyrosine kinase substrate, Hrs). These proteins signal for entry into the MVB pathway, a process regulated by the binding of vps27p to PtdIns(3)P, ubiquitin and the ESCRT-1 complex.[214] SARA regulates the subcellular localisation of R-Smads and presents them to the TGF-β receptor. Localisation of SARA in early endosomes is required for efficient TGF-β/Smad signalling.[215]

8.2. PtdIns(3,5)P_2

In yeast, PtdIns(3,4)P_2 maintains vacuole membrane integrity and turnover.[216] Its levels increase dramatically upon hyper-osmotic shock in yeast and UV irradiation of mammalian cells.[14,217] Whilst the motifs that bind PtdIns(3,5)P_2 are poorly characterised but the Epsin N-terminal homology (ENTH) domain of ent3p binds PtdIns(3,5)P_2 and is required for entry into the lumen of the multivesicular body.[208,218] A yeast protein Svp1 (also known as ATG18/AUT10/CVT18) binds PtdIns(3,5)P_2 with high specificity and affinity.[219] Svp1p plays a role in membrane recycling from the vacuole. In humans, there are at least four orthologues but only WIPI-49 has been studied which binds PtdIns(3)P, PtdIns(3,5)P_2 and PtdIns(5)P.[220]

8.3. PtdIns(3,4)P_2 and PtdIns(3,4,5)P_3

PH and PX domains that display a binding specificity for PtdIns(3,4,5)P_3 and PtdIns(3,4)P_2 are found in a number of intracellular signalling molecules. These include serine/threonine protein kinases, tyrosine kinases, phospholipases and molecules that regulate the exchange of guanine nucleotides to GTPases.

8.3.1. *Serine/threonine protein kinases*

It is clear that production of PtdIns(3,4)P_2 and PtdIns(3,4,5)P_3 results in the phosphorylation of a number of serine/threonine protein kinases.

8.3.1.1. Akt/PKB

Akt (also known as protein kinase B-PKB and RAC related to protein kinase A and C) was identified as the cellular homologue of the viral oncoprotein v-Akt[221] and as a homologue of protein kinase A and protein kinase C.[222,223] There are three isoforms of the 57 kDa Akt serine/threonine kinase termed Akt1, Akt2 and Akt3 and each are encoded by separate genes on chromosomes 14q32, 19q13 and 1q44, respectively to constitute a gene family.[223–227] Akt2 and Akt3 show 81% and 83% sequence similarity with Akt1, respectively and are thought to be activated similarly[228] (Fig. 7). All isoforms have a wide tissue distribution although Akt1 is highly

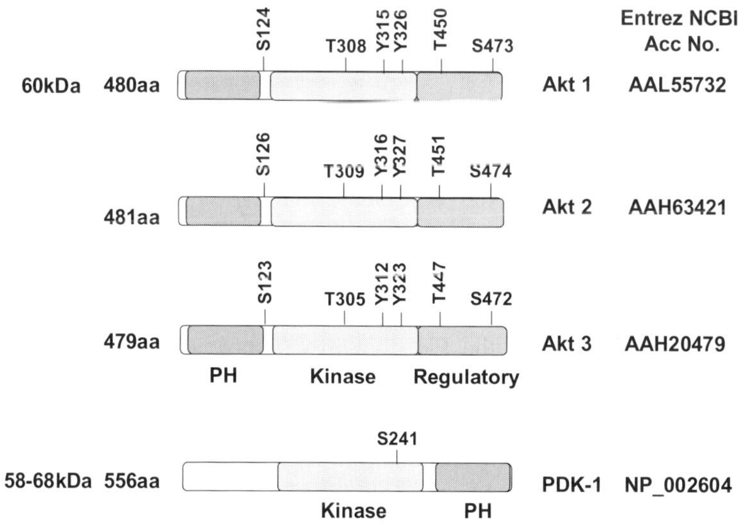

Fig. 7 Schematic representation of the Akt/protein kinase B (PKB) family and phosphoinositide-dependent kinase-1. Akt isoforms Akt1, Akt2 and Akt3 contain an N-terminal pleckstrin homology domain (PH), a catalytic domain (kinase) and a C-terminal regulatory domain (regulatory). Equivalent sites of phosphorylation on threonine (T) and serine (S) residues are indicated.

expressed in the brain, heart and lung, Akt2 is expressed largely in the skeletal muscle and embryonic brown fat, and Akt3 is expressed in the brain, kidney and embryonic heart.[227,229,230] Mice null for either Akt1 or Akt2 are viable and have relatively minor phenotypes suggesting functional redundancy amongst this kinase family. The expression of Akt is observed in all eukaryotes but is absent in yeast.

8.3.1.2. Akt structure and regulation of kinase activity

All Akt isoforms contain an N-terminal PH domain, a central catalytic domain and a C-terminal regulatory region. Growth factor and cytokine-mediated generation of PtdIns(3,4,5)P_3 and PtdIns(3,4)P_2 results in the translocation of Akt to the plasma membrane and the binding of these 3-phosphoinositides via the PH domain.[231] Mutations within the PH domain of Akt block activation of catalytic activity *in vivo*[232] and by 3-phosphoinositides *in vitro*.[196] Akt is constitutively phosphorylated on residues serine 124 and threonine 450 but its ligand induced phosphorylation on threonine 308 within the kinase domain and serine 473 at the C-terminus confers maximal specific activity.[233] Importantly, phosphorylation of both threonine 308 and serine 473 and translocation of Akt to the membrane following PDGF stimulation is inhibited by wortmannin. Threonine 308 lies within the kinase domain-activating loop. Akt immunoprecipitated from serum-starved cells is inactive but its specific activity is rapidly increased following serum stimulation.[232,234] The increased Akt kinase activity by PDGF was dependent upon Y740 and Y751, the same phosphotyrosine residues involved in binding the class IA PI3K adaptor p85. Akt mutated in the PH domain was not activated by PDGF.[232] The Akt PH domain binds both PtdIns(3,4,5)P_3 and PtdIns(3,4)P_2. Neither PtdIns(3)P or PtdIns(3,5)P2 binds this PH domain.[235] The role of the PH domain in Akt activation has been highlighted by the development of phosphatidylinositol ether lipid analogues (PIAs) that bind the Akt PH domain (IC50 < 5 mM).[236]

Although phosphorylation of serine 124 and threonine 450 does not regulate kinase activity but may be required for the interaction of Akt with heat shock protein (Hsp90) that mediates protein folding. Stimulation of the Src tyrosine kinase-dependent activation of Cdc42 downstream of integrins activates Akt.[237] Cell stress induced by increased reactive oxygen species

(ROS) or the accumulation of cell metabolites in the extracellular medium increases Akt phosphorylation.[238,239] Following growth factor receptor activation, Akt translocates to the nucleus.[240] Akt activity is negatively regulated by CTMP, a protein that binds the C-terminal regulatory domain of Akt inhibiting the phosphorylation of threonine 308 and serine 473.[241]

8.3.1.3. Phosphoinositide-dependent kinase — PDK-1

PDK-1 was isolated from bovine brain and rabbit skeletal muscle using biochemical purification strategies to isolate the kinase responsible for the phosphorylation of Akt on residue Thr 308.[235,242] This serine/threonine kinase contains an N-terminal catalytic domain with sequence similarity to PKA, Akt and PKC. At the C-terminus is a PH domain that binds PtdIns(3,4,5)P_3 with high affinity.[243] Phosphorylation of Akt by PDK1 is dependent on 3-phosphoinositide production. Akt mutants that lack its PH domain indicate that 3-phosphoinositide binding plays a regulatory role upon PDK-1 phosphorylation relieving the inhibitory effects of the PH domain. Phosphorylation of Akt on threonine 308 and serine 473 occurs at the plasma membrane to which both Akt and PDK1 translocates following receptor activation. Constitutive localisation of Akt at the plasma membrane results in its constitutive phosphorylation and activation.[244] A PI3K-dependent translocation of PDK1 to the nucleus has been also described.[245] The nuclear export signal (NES) in mouse PDK-1 has been found at amino acid residues 382–391, and it has been proposed that nuclear translocation may be an inhibitory mechanism to sequestrate PDK-1 from cytosolic signalling pathways. In keeping with its regulatory effects on the PI3K-Akt signalling pathway, loss of PDK1 expression arrests T cell differentiation in the thymus, whereas reduced expression allowed T cell differentiation but blocked their proliferative expansion.[246]

8.3.1.4. Additional downstream targets of PDK-1

Both the threonine 308 and serine 473 residues of Akt lie within consensus sequences shared by the AGC family of protein kinases that include protein kinases A, G and C.[247] Other proteins include p70S6-kinases (p70S6K), p90 ribosomal S6-kinases (p90RSK) and serum- and glucocorticoid-induced protein kinase (SGK).[248] As for Akt, phosphorylation of these proteins

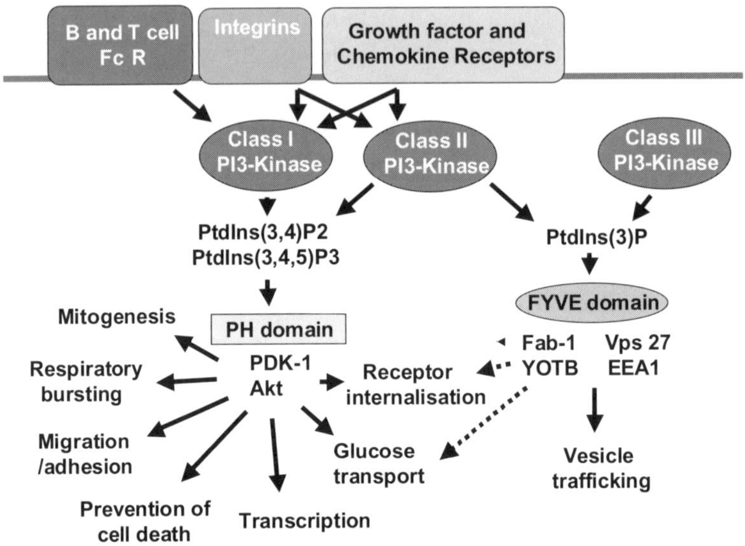

Fig. 8 Regulation of PIK isoforms and activation of cellular effector pathways. Growth factor and chemokine receptors, integrins, B-cell, T-cell and Fc receptors translocate and activate class I and class II PI3K enzymes to produce PtdIns(3,4,5)P_3 and potentially PtdIns(3,4)P_2. Each of these 3-phosphoinositide products serve to translocate phosphoinositide-dependent kinase-1 (PDK-1) and Akt via binding to their pleckstrin homology domain (PH). Phosphorylation of Akt and other proteins by PDK-1 stimulates a host of cellular effects. Receptor-mediated activation of class II PI3K and class III PI3K enzyme activity generates PtdIns(3)P that translocates FYVE domain-containing proteins that include Fab-1, vps27, YOTB and early endosomal antigen-1 (EEA-1) that in turn regulate vesicle traffic, receptor internalisation and movement of the glucose transporter.

in vivo increases following receptor stimulation and it is dependent on 3-phosphoinositide production (Fig. 8). Consequently, several of these kinases become phosphorylated by PDK-1 on a residue equivalent to threonine 308.[249–253] It is important to note that PKA phosphorylates BAD on residue serine 112 thereby providing a second mechanism to prevent cell death.[254]

P70S6K exists in two isoforms termed p70S6Kα and p70S6Kβ. The localisation of p70S6Kβ is primarily nuclear whereas p70S6Kα is mainly found in the cytoplasm. The activity of this kinase can be inhibited by rapamycin or indirectly by wortmannin.[255,256] p70S6K is involved in G1 to S phase cell cycle progression and the translocation of mRNA messages that contain a polypyrimidine tract at their translational start site.[257] These genes include transcripts for ribosomal proteins and elongation factors of protein synthesis. The main substrate of p70S6K protein kinase activity is the S6 ribosomal protein that comprises the 40S ribosomal subunit. The Rho family GTPases Rac1 and Cdc42 bind p70S6K and promote its phosphorylation.[258]

The family of PKC enzymes regulate protein synthesis, transcription, cell growth, differentiation and apoptosis.[259] The kinase activity of several PKC isoforms is directly activated by 3-phosphoinositides *in vitro* and inhibited *in vivo* by PI3K inhibitors wortmannin or LY294002.[260–262] Phosphorylation of PKCδ and PKCζ by PDK1 is stimulated by PtdIns(3,4,5)P_3 *in vitro*.[263]

cAMP-dependent protein kinase (PKA) is a complex kinase family consisting of at least three catalytic and two regulatory subunits.[264] PKA is activated by cAMP produced downstream of G-protein coupled receptors. Once activated, PKA phosphorylates a number of downstream targets that regulate cell growth, differentiation and metabolism.[265] PDK-1 mediated phosphorylation of PKA on residue threonine 197 increases kinase activity.[249,266]

8.3.1.5. Integrin-linked kinase (ILK)

Following the isolation of PDK-1 as the kinase responsible for phosphorylating Akt on residue threonine 308, several kinases were proposed as the PDK-2 enzyme that phosphorylates Akt on residue serine 473. These include MAPKAP kinase-2, several PKC isoforms, integrin-linked kinase (ILK), DNA-dependent protein kinase, Ataxia-telangiectasia mutant (ATM) and also PDK-1.[267] Others have suggested that PDK-2 may not exist and that Akt/PKB autophosphorylates itself on serine 473.[268] Using double-stranded RNA interference (siRNA) and conditional knockouts of ILK show that loss of ILK does not affect phosphorylation of Akt on threonine 308 but does abolish phosphorylation of serine 473, suppressed phosphorylation of GSK-3β and expression of cyclin D1 and simulated apoptosis.[269]

The mammalian target of rapamycin (mTOR) in complex with proteins raptor and GβL phosphorylates several proteins in a hydrophobic motif equivalent to that found in Akt and which contains serine 473.[270,271] However, since phosphorylation of this residue is not sensitive to rapamycin, this possibility was not actively pursued. Recently however, a rapamycin-insensitive form of the mTOR protein kinase in which it exists as a complex with Rictor:GβL has been shown to mediate phosphorylation of Akt/PKB at serine 473.[272] Knockdown of mTOR or Rictor expression by RNAi ablated serine 473 phosphorylation in mammalian and *Drosophila* cells. Also, an mTOR:Rictor:GβL complex efficiently phosphorylated Akt/PKB at Ser473 *in vitro*. Finally, the serine 473 Akt kinase activity of the mTOR:Rictor:GβL complex was stimulated by serum *in vivo*. A PH domain leucine-rich repeat protein phosphatase (PHLPP) was identified as a serine 473 protein phosphatase. PHLPP belongs to a poorly studied okadaic acid-insensitive class of protein phosphatase termed 'PP2C' (or PPM family), of which 12 members are encoded in the human genome.[273]

8.3.1.6. Downstream targets of Akt

Disruption of Akt1 and Akt2 produces viable mice with contrasting phenotypes. Akt1 knockout mice are small and have a shorter life expectancy than wild-type littermates under conditions of stress.[274] Placental abnormalities result in intrauterine growth retardation.[275] These animals also have increased apoptotic cell death in thymus and testis and impaired platelet responses to thrombin and collagen.[276] Conversely, knockout of Akt2 produces insulin resistance and a state similar to diabetes mellitus.[277] The relatively mild phenotypes suggest functional redundancy thus making data interpretation difficult.

A wide spectrum of proteins present in both the cytoplasm and nucleus has been found to act as substrates of Akt kinase activity. These are divided into a number of functional groups (Fig. 9).

8.3.2. *Prevention of cell death*

The anti-apoptotic activity of Akt was the first and remains the most studied regulatory function of this enzyme. Akt is frequently over-expressed in a number of tumours[278] and expression of the recombinant enzyme delays cell death.[279] At the molecular level, several molecules have been identified as substrates of Akt kinase activity and whose phosphorylation regulates cell death.

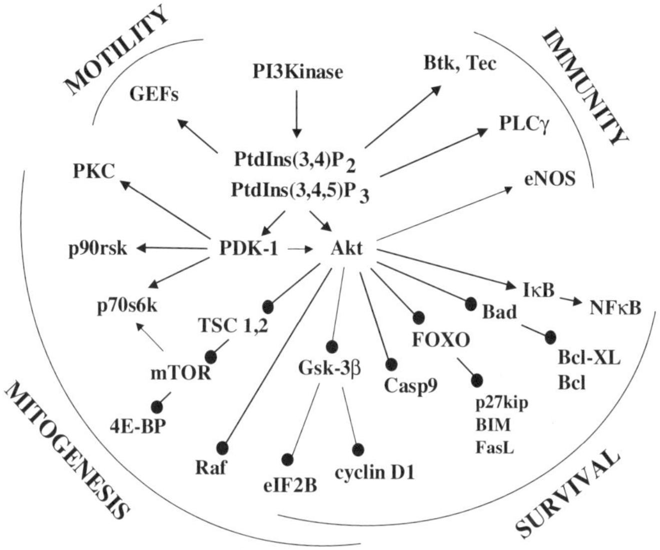

Fig. 9 Effectors of PtdIns(3,4)P_2 and PtdIns(3,4,5)P_3 signalling. Targets of each 3-phosphoinositide or the serine/threonine kinases phosphoinositide-dependent kinase-1 (PDK-1) and Akt are shown. Arrows indicate stimulation, filled circles indicate an inhibitory effect.

In its non-phosphorylated form, the Bcl-2/Bcl-XL antagonist causing cell death (BAD) protein is localised in mitochondria where it forms a complex with two anti-apoptotic proteins termed Bcl-2 and Bcl-XL thereby inhibiting their activity.[280] Phosphorylation of BAD on either serine 112 or serine 136 prevents its association with Bcl-2 and Bcl-XL leading to its release into the cytoplasm where it binds 14-3-3 proteins.[281] Akt phosphorylates BAD on serine 136.[282,283]

Caspase 9, a pro-apoptotic protease is phosphorylated and inhibited by Akt.[284] Akt also phosphorylates p21(Waf1/Cip1) and p27 (Kip1) and inhibits their anti-proliferative effects by retaining them in the cytoplasm.[285–287] In this way, the activation of cyclin/Cdk complexes such as cyclin D1/Cdk4 is inhibited. Akt also phosphorylates the murine double minute-2 (mdm-2) protein leading to its translocation into the nucleus and destabilising the tumour suppressor protein p53.[288,289] Degradation of p53 inhibits the stress response thereby increasing cell survival.

8.3.3. *Regulators of transcription*

The forkhead box O (FOXO) family of transcription factors are direct targets of Akt.[290,291] FOXO1 (FKHR), FOXO3a (FKHRL1), FOXO4 (AFX) and FOXO6 are predominantly localised in the nucleus of quiescent cells.[292] Following cell stimulation, FOXO proteins translocate to the cytosol where they interact with 14-3-3 which sequesters them away from the nucleus.[293] Phosphorylation of FOXO proteins by Akt occurs on three highly conserved consensus motifs (RxRxxS/T) to regulate this process.[291,294,295] In doing so, it prevents the expression of the pro-apoptotic Fas ligand IGFBP-1 and Bim. Another kinase that phosphorylates FOXO factors is serum and glucocorticoid–regulated kinase (SGK).[296] SGK is activated by 3-phosphoinisitides and PDK-1 but because it lacks a PH domain it is not recruited to the plasma membrane following receptor activation. *De novo* expression of SGK is rapidly induced following serum, glucocorticoid stimulation and stress, and both Akt and SGK phosphorylate the same residues making their relative contribution to the regulation of FOXO activity difficult to distinguish.[292,296] Akt also phosphorylates and activates IkappaB kinase a which phosphorylates IkappaB targeting it for degradation.[297,298] This results in nuclear translocation and activation of nuclear factor-kappaB (NF-kappaB) and transcription of NF-kappaB dependent pro-survival genes such as Bcl-XL, caspase inhibitors and c-Myb.[299] Akt phosphorylates and activates cyclic AMP response element binding protein which increases the transcription of anti-apoptotic genes that include Bcl-2, Mcl-1 and Akt itself.[300–302] In tumour cell lines a lack functional PTEN results in the constitutive activation of FOXO proteins.[303] FOXO proteins are considered not only to be critical effectors of PTEN regulated cell cycle arrest and apoptosis but also effectors of FasL.[304] In *C. elegans* the FOXO orthologue Daf16 lies downstream of a PI3K/Akt pathway.[305]

Nuclear Factor κB (NFκB) is a transcription factor for several anti-apoptotic genes.[306] When bound to its inhibitor IkB, NFκB is inactive and retained in the cytoplasm Akt binds and activates an IkB kinase (IKK).[307] Phosphorylation of IkB by IKK results in the dissociation of this complex allowing NFκB to move to the nucleus and targets IkB for degradation in the proteosome.

PI3K enzymes and Akt also regulate the activity of the c-fos promoter in HeLa cells downstream of Rho- and Rac-dependent signals through the serum response element (SRE).[308,309] In addition, Akt directly phosphorylates CREB on Ser 133.[300,310] Phosphorylated CREB binds the co-activator CREB binding protein (CBP) that promotes expression of genes that suppress apoptosis.[311]

8.3.4. *Regulators of metabolism*

Following stimulation of the insulin receptor, Akt directly phosphorylates and inactivates glycogen synthase kinase 3 (GSK3) on serine 9.[312,313] GSK3 was the first documented target of the Akt kinase. Inhibition of GSK3 by Akt prevents the phosphorylation of β-catenin which inhibits its degradation and it translocates to the nucleus. There β-catenin binds transcription factors like TCF/LEF-1 to induce expression of genes such as cyclin D1. GSK-3 phosphorylates and inactivates the translational initiation factor eIF2B.[314] Since GSK3 is inactivated by growth factor stimulation, this could explain their effects on translation initiation.[315] However, treatment of cells with compounds that directly inhibit GSK3 leads to the dephosphorylation of this site, but does not activate eIF2B suggesting greater complexity.[316] Akt also phosphorylates and activates 6-phosphofructo-2-kinase (PFK-2) an enzyme responsible for synthesis of fructose 2,6-biphosphate.[317] Fructose 2,6-biphosphate in turn stimulates 6-phosphofructo-1-kinase, a potent activator of glycolysis.[318]

The growth of eukaryotic cells is regulated not only by intercellular growth factors but also by the availability of nutrients.[319–321] The Akt sub strate tuberous sclerosis complex-2 (TSC2) forms part of the TSC1/TSC2 protein complex constituting a nutrient-sensitive signalling pathway that alters cell size and growth.[322,323] Phosphorylated TSC2 fails to bind TSC1 which inhibits p70S6K and activates 4E-BP1. Germ line mutations in either tumour suppressor genes TSC1 or TSC2 result in tuberous sclerosis, an autosomal dominant human genetic disorder characterised by hamartoma development.[324] The TSC1/TSC2 complex has GAP activity for the Rheb GTPase (a member of the ras family), and activated Rheb-GTP activates the mammalian target of rapamycin (mTOR).[185,325] mTOR promotes translation of cyclin D mRNA and p70 S6K while the eukaryotic initiation factor 4E-binding protein 1 (4E-BP1) is inhibited. Using a mouse model of prostate cancer, pharmacological inhibition of mTOR by the rapamycin analogue RAD001 eradicated the prostate intraepithelial neoclassic (PIN).[326]

8.3.5. *Protein tyrosine kinases*

The Tec kinase family of non-receptor tyrosine kinases are expressed in haematopoietic cells.[99,327] They contain N-terminal SH2, SH3 and PH domains with the PH domain being responsible for membrane localisation of the Tec kinases.[328,329] Germ line mutations in the Bruton tyrosine kinase (Btk) whose PH domain binds PtdIns(3,4)P_2 and PtdIns(3,4,5)P_3 with high affinity, leads to defective B-cell maturation and agammaglobulinaemia.[330] Several of these mutations map to the PH domain of Btk highlighting the biological significance of this domain.[331] Two mice strains have a similar phenotype; Xid mice that have a mutation of Btk in their PH domain[332] resulting in defective binding of its mutated PH domain to PtdIns(3,4,5)P_3,[333] and secondly mice with a targeted disruption of the class IA PI3K adaptor subunit p85α.[334] Such data indicates that PI3K signalling pathway is functionally linked to the Btk pathway in antigen receptor-mediated signal transduction and is pivotal in B-cell development and functions. B-cell receptor aggregation-driven CD19 phosphorylation functions to promote Btk activation via the translocation of PI3K enzymes. The resultant production of PtdIns(3,4,5)P_3 localises Btk for subsequent phosphorylation and activation by Src and Syk family kinases.[335]

8.3.6. *Phospholipases*

Phospholipase C hydrolyses phospholipids to produce diacylglycerol (DAG) and inositol trisphosphate (IP3).[336] DAG activates protein kinase C whilst IP3 mobilises the release stores of intracellular calcium ions.[337] The isoform phospholipase Cγ2 (PLCγ2) is a target of Btk and it has a PtdIns(3,4,5)P_3-specific PH domain that mediates the translocation of this PLC isoform to the plasma membrane.[338]

8.3.7. *GTP exchange factors (GEFs) and GTPase activating proteins (GAPs)*

Small GTPases alternate between a GTP-bound 'active' state and a GDP-bound 'inactive' state.[339] Interchange between these states is regulated by GTP/GDP exchange factors (GEFs) and GTPase activating proteins (GAPs). GEFs catalyse the exchange of GDP for GTP. In contrast, GAPs accelerate the intrinsic GTPase activity. GEFs of the Rho GTPase family that includes Rac, Rho and Cdc42 and the GEF for Arf contain a PH domain.[340] The family of Rho GEFs has over 60 isoforms characterised by a ~200 residue Dbl homology (DH) domain followed almost invariably by a PH domain. Outside of the characteristic DH and PH domains, family members are highly diverse in sequence. Treatment of cells with PI3K inhibitors inhibits receptor-mediated induced reorganisation of the actin cytoskeleton and Rac GTP binding.[341] This effect can be reversed by expression of constitutively activated Rac, indicating that it lies downstream of 3-phosphoinositide production to regulate actin filaments. Vav and Tiam1 are two Rac GEFs that may potentially mediate this effect.[342,343] Although ARF GTPases are involved in intracellular vesicle transport, three ARF GEFs termed General Receptor for Phosphoinositides (GRP1), ARF nucleotide binding site opener (ARNO) and cytohesin-1 contain PH domains that preferentially bind PtdIns(3,4,5)P_3 allowing receptor-stimulated translocation of these GEFs from the cytoplasm to the plasma membrane.[344] $GAP1^m$ and GAP1IP4BP are GAPs for Ras and their binding to the plasma membrane is regulated by PtdIns(3,4,5)P_3.[345] Centaurins also bind PtdIns(3,4,5)P_3 *in vivo*[346] and have homology to the ARF GAP in yeast.

9. Manipulation of PI3K Enzyme Expression

9.1. Manipulation of PI3K expression *in vitro*

Although transient expression of class I PI3K enzymes produces a marked increase in cellular levels of PtdIns(3,4,5)P_3, stable over-expression of class I PI3K enzymes has proved unsuccessful to date due to toxicity. Over-expression of the class IA adaptor or a mutant that is unable to bind the p110 catalytic subunit (Δp85) have been used to provide a dominant negative effect.[347,348] Microinjection of antibody to class IA PI3K isoforms has been used with great effect to demonstrate the role of p110α in PDGF and CSF-1 stimulated proliferation and PDGF stimulated actin reorganisation.[349] In contrast, antisera against p110β inhibited LPA and insulin induced mitogenesis.[107,350] In microinjection of antibody revealed the role of p110β and p110δ in regulating the actin cytoskeleton of macrophages.[351] More recently, an increasingly used technique to attenuate protein expression is RNAi.[352]

9.2. Manipulation of PI3K expression *in vivo*

The embryonic development of mice lacking either all isoforms of the p85α gene (p85α−/−p55α−/−p50α−/−) the p85β gene (p85β−/−) is not impaired.[353,354] Loss of p85α and all of its possible splice variants results in death shortly after birth, whilst deletion of p85β produced no phenotype.[355] Tissue-specific deletion in lymphocytes reveals that deletion of p85 impaired B-cell development and function but T-cells were normal.[356]

Deletion of the p110α and p110β catalytic subunit is embryonically lethal with embryos developmentally delayed at embryonic day (E) 9.5 and dead between E9.5 and E10.5.[357,358] In these cells, expression of p85α and p55 is markedly increased and may exert a dominant negative effect on other class IA PI3K isoforms contributing to the lethality. Heterozygous p110α+/−p110β+/− null mice exhibit mild glucose intolerance and hyperinsulinemia in the fasted state.[359] In addition, these mice also show an approximately 50% decrease in p85 expression in liver and muscle. RNA interference has been used to attenuate the expression of class I PI3K in *Drosophila*[360,361] and mammalian cells.[357,362] In human cells, short hairpin RNA (shRNA) against p110α attenuated the expression of p110α without affecting p110β, p110δ, or p85α.[357] In a separate study knockdown of p110 by RNAi resulted in loss of p85 proteins due to decreased protein stability.[359]

In contrast, p110γ and p110δ knockout mice grow normally[363–369]; however in mice that lack p110δ B-cell development is impaired at an early pre-B stage[365] generation of marginal zone B-cells and germinal centre formation are blocked and *in vivo* antibody production in response to both T-dependent and -independent antigen is severely impaired.[364] Since p85 null mice have defects only in the T-independent antibody response,[370] this suggests that p85 may utilise additional adaptor proteins. Loss of p110δ in mast cells leads to defective proliferation, adhesion and migration and impaired response to allergen.[371] Mice null for p110γ have altered macrophage and neutrophil function.[367,372,373] Besides effects on neutrophil migration, these animals show a reduced ability to produce reactive oxygen species and reduced inflammatory responses to microbial infection.

10. Biological Role of PI3K Signalling Pathway

10.1. Vesicle transport

Analysis of mutant yeast strains demonstrated that vps34p is required for the transport of newly synthesised proteins from the Golgi to the vacuole which is equivalent to the lysosome in mammals. In mammalian cells, the use of PI3K inhibitors and expression of vps34p orthologues demonstrated that vps34p plays a direct role in targeting lysosomal precursors to the endocytic pathway in an analogous fashion to its role in the fusion of early endocytic vesicles with endosomes in yeast.[374] Production of PtdIns(3)P translocates FYVE domain-containing proteins that play a role in the endocytic and secretory pathways. In mammals, fusion of early endosomes is regulated by the small GTPase Rab5 and requires the recruitment of the early endosomal protein EEA1 by PtdIns(3)P.[155,188,375] Insulin regulates glucose uptake by stimulating translocation of the GLUT4 transporters to the cell surface.[376] The return of GLUT4 from the plasma membrane to insulin-sensitive store is also sensitive to wortmannin.[377] The involvement of vps34p in this process has been proposed on the basis of the high sensitivity of human vps34p to wortmannin[378] and the role of vps34p in post-endocytic receptor sorting and fusion of early endosomes.[379] Expression of myotubularins, that act as a PtdIns(3)P phosphatase, impairs insulin-stimulated glucose uptake in 3T3-L1 adipocytes.[380] Also, insulin-stimulated GLUT4 translocation in 3T3-L1 cells is blocked by antibody raised against the PtdIns(3)P binding protein sorting nexin 9.[381]

During periods of extended nutrient deprivation, cytoplasm and organelles are sequestered into double or multilamellar autophagic vesicles.[382] These autophagosomes are then delivered to lysosomes for degradation. Autophagy plays a role in tumour suppression, protein degradation and organelle turnover.[383] In vps34p mutant temperature-sensitive yeast strains, macroautophagy is inhibited and in mammalian cells macroautophagy is inhibited by wortmannin and stimulated with PtdIns(3)P.[160,161,384,385] In yeast, the vps34p/vps15p heterodimer has been found in complex with Agp6 and Agp14 two proteins that regulate autophagy.[157,386,387]

Although most emphasis has been placed on the role of the class III PI3K enzymes and PtdIns(3)P in the regulation of vesicle transport, it is clear that other PI3K enzymes play a regulatory role. Inhibition of PI3K activity inhibits insulin-stimulated glucose transport,[348] and translocation of the GLUT4 glucose transporter to the plasma membrane in response to insulin involves vps34p-independent production of PtdIns(3)P.[388] Expression of the class II PI3K PI3K-C2α alters the endocytosis of transferrin.[138,389] Similarly, injection of antisera to the classIA PI3K p110α inhibited transferrin recycling.[379,390] The class IA heterodimers p85/p110β associate with and activate Rab5 a small GTPase that regulates fusion of early endosomes,[391,392] and PtdIns(3,4,5)P_3 regulates the activity of the ARF GTPase responsible for endosome fusion.[393]

10.2. Receptor downregulation

Following growth factor receptor activation and the resultant tyrosine kinase activity, the signalling complex is internalised to regulate cell responsiveness and provide temporal and spatial specificity of the downstream signalling events. The process of receptor internalisation into intracellular vesicles and its degradation in lysosomes is regulated by 3-phosphoinositide generation.[394,395] Point mutation of the p85 binding site on the PDGF receptor and the PI3K inhibitor wortmannin abolish the transfer of receptor to lysosomes.[394,396]

10.3. Cell viability and longevity

In *C. elegans*, the dauer phenotype is a larval state characterised by developmental arrest and decreased metabolism triggered by adverse environmental conditions such as nutrient deprivation. Its analysis has identified a number of dauer affected (daf) genes, some of which are orthologues of the PI3K signalling pathway.[397,398] Age1 is an orthologue of p110α. In *Drosophila* a loss of PTEN at the larval stage is lethal. This phenotype can be rescued by expression of Akt.[399] *In vitro*, both LY294002 and the MEK-1 inhibitor PD58029 arrest fibroblast proliferation. However, only LY294002 induces effects consistent with ageing such as galactosidase activation, collagenase gene over-expression and decreased expression of early population doubling level cDNA (EPC-1).[400]

10.4. Mitogenesis

Stimulation of PI3K activity occurs in response to treatment of cells with mitogens. 3-phosphoinositide production seems essential for cell cycle entry and the rate of cell cycle progression. The role of PI3K activity in cell proliferation has been demonstrated in several ways that include the use of pharmacological inhibitors, microinjection of the SH2 domain of the class IA adaptor, neutralising antibodies and mutagenesis of receptor phosphotyrosine binding sites.[401–403] Given their association with transforming oncogenes, the translocation of classIA PI3K enzymes to the plasma membrane and the generation of PtdIns(3,4,5)P_3 following growth factor stimulation, most attention has focused on them as mediators of cell proliferation. However, the intricate relationship between the PI3K isoforms was illustrated by the observation that neutralising antibody to vps34 blocks cell proliferation downstream of the insulin receptor and is required at a different stage of the G1-S phase transition of the cell cycle compared to class IA PI3K enzymes.[379,404]

10.5. Cell volume

Expression of the class I PI3K enzyme in *Drosophila* increased cell growth and organ size.[405] Correspondingly, mutations in the insulin signalling pathway in *Drosophila* regulates cell size and number.[406] These effects are thought to result from a consequence of altering the activity of the mTOR orthologue. Neurones lacking the 3-phosphoinositide phosphatase PTEN have elevated levels of Akt phosphorylation and a corresponding increase in soma size without a corresponding increase in the rate of cell proliferation.[407]

10.6. Differentiation

A role for the PI3K signalling pathway in differentiation has been demonstrated for many cell types.[408–411] PC12 is a pheochromocytoma cell line that has been used extensively to study cell differentiation upon addition of growth factors that include nerve growth factor (NGF).[412] Activation of the PI3K-Akt pathway downstream on the NGF receptor promotes PC12 cell survival[33] and it is essential for NGF induced differentiation.[410] PI3K activity is also required for differentiation of fetal brown adipocytes and muscle progenitor cells[413,414] In myoblasts, inhibition of the Raf-MEK-Erk pathway promotes muscle differentiation whereas inhibition of PI3K enzyme activity blocks differentiation.[408,415] The degree to which a stimulus activates each of these signalling pathways appears critical for the resultant phenotype. Increased Akt activity also inhibits the Raf-MEK-Erk signalling pathway at the level of Raf or above but only in differentiated myotubes not myoblasts. In mast cells, a dependence of the Raf-MEK-Erk signalling pathway on PI3K activation has also been demonstrated, however in the haematopoietic system this regulation only occurs in immature haematopoietic progenitor/stem cells.[416] In this way, the role of the PI3K-Akt signalling pathway differs depending upon the stage of cell differentiation. Paradoxically, inhibition of PI3K activity in human foetal undifferentiated cells stimulated endocrine differentiation.[417] In addition, the proportion of pluripotent precursor cells increased together with the number of resultant differentiated cells.

Treatment of the human leukaemia cell line U937 with all-trans retinoic acid (ATRA) induces their differentiation to macrophage-like cells and increased their expression of p110γ.[418] In contrast, expression of class IA PI3K enzymes and p101 were unaltered. The treatment of HL-60 cells with ATRA induces their differentiation to granulocytes and elevates the specific activity of class IA PI3K activity.[419] In addition ATRA also increases the specific activity, tyrosine phosphorylation and nuclear translocation of the class II PI3K isoform PI3K-C2β in these cells.[420]

10.7. Migration

The role of PI3K enzymes and PtdIns(3,4,5)P_3 in regulating cell migration has been demonstrated extensively.[421,422] The effect of PI3K activation on the actin cytoskeleton are mediated through Rac and Cdc42 leading to the reorganisation of actin filaments, increased motility and invasiveness. In mouse embryo fibroblasts lacking all p85α and p85β gene products, PDGF cannot stimulate membrane ruffling.[355,423] Re-introduction of either p85α, p85β or p50α rescues the membrane ruffling defect demonstrating that the SH3 domain and the Rho/Rac/Cdc42-interacting domain of p85 are not required for this response.

11. Phosphoinositide Phosphatases

The concerted action of a set of phosphatases that include PTEN, myotubularin and SHIP serve to negatively regulate PI3K signalling (Fig. 10).

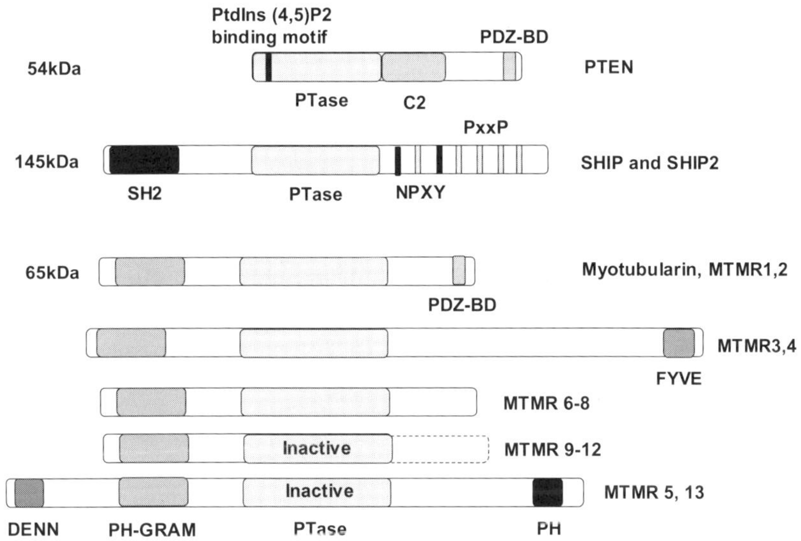

Fig. 10 Schematic representation of 3- and 5-phosphoinositide phosphatases. The N-terminus of the 3-phosphoinositide phosphatase PTEN is formed by a catalytic phosphatase domain (PTase) that contains a PtdIns(4,5)P2-binding motif. Adjacent to this lies a phospholipid-binding C2 domain with a **PS**D-95, **d**iscs large and **Z**O-1 binding domain (PDZ-BD) at the C-terminus. The 5-phosphoinositide phosphatases SHIP and SHIP2 contain an N-terminal src homology 2 domain (SH2), the phosphatase domain (PTase) lies towards the C-terminus followed by two NPXY motifs that when phosphorylated on tyrosine potentially bind phosphotyrosine-binding (PTB) or SH2 domain-containing proteins. The C-terminus contains many polyproline-rich (PxxP) domains that are a consensus for SH3 domain-binding. The myotubularin family of proteins are PtdIns(3)P hydrolysing 3-phosphatases. Myotubularin and myotubularin-related (MTMR) phosphatases all contain an N-terminal phox homology — Glucosyltransferase, Rab-like GTPase activator and myotubularins domain (PH-GRAM), and a phosphatase domain (PTase) which in MTMR5 and MTMR9–12 is catalytically inactive. Myotubularin and MTMR1 and 2 contain a PDZ-binding domain (PDZ-BD) at the C-terminus, MTMR3 and 4 have a FYVE domain in this position and MTMR5 and 13 has a PH domain. Both MTMR5 and 13 also have an N-terminal differentially expressed in neoplastic versus normal cells domain (DENN) that is formed by three sub-domains and is involved in Rab-mediated processes or regulation of MAPK signalling pathways.

11.1. PTEN

PTEN (**p**hosphatase and **ten**sin homologue deleted on chromosome 10) also known as MMAC1 (**m**utated in **m**ultiple **a**dvanced **c**ancers) was characterised by two groups as a candidate tumour suppressor gene localised to chromosome 10q23.[424,425] Deletions in 10q 22–25 occur in multiple tumour types including glial, prostate, endometrial, small cell lung cancer and renal cell cancer. In addition, the gene was also identified by a group working independently to identify protein phosphatases and named it TGF-β regulated and epithelial cell enriched phosphatase (TEP-1).[426] The sequence of the PTEN gene product indicated that it was a member of the protein tyrosine phosphatase (PTP) superfamily of enzymes. These PTPs regulate pivotal cellular functions including cell growth, metabolism, differentiation, motility and cell death and all contain the signature motif Cys-X5-Arg (CX5R). Specifically, the PTEN gene sequence indicated that it might act as a dual specificity phosphatase able to act on phosphotyrosine, phosphoserine and phosphotyrosine residues.

11.1.1. *PTEN as a tumour suppressor gene*

Mutations in the PTEN gene occur at high frequencies in human tumours, xenografts and cancer cell lines indicating a close correlation between the loss of PTEN function and malignancy.[424,425] Germ line mutations in the PTEN gene are associated with Cowden syndrome, Bannayan-Zonanan syndrome, Lhermitte-Duclos disease and juvenile polyposis coli. Over 80% of patients with Cowden syndrome and a similar frequency of patients with Lhermite-Duclos disease carry germ line mutations in PTEN.[427,428] These syndromes are characterised by the development of hamartomas, hyperplastic disorganised benign tumours. Increased proliferation of hamartomas is observed upon the loss of the wild type PTEN allele. Patients with Cowdens disease are pre-disposed to tumours including breast and thyroid.[429] Lhermitte-Duclos disease is defined by a dysplastic gangliocytoma of the cerebellum, a pathology observed in mice with a deletion of PTEN in the brain.[430] Germ line mutations in PTEN that give rise to these syndromes comprise nonsense, missense frameshift and splice site mutations, resulting in a truncated or inactive enzyme. Over expression of PTEN in glioma cells that contain a mutated *PTEN* gene results in growth suppression,[431] and finally mice heterozygous for PTEN gene develop hyperplasia and dysplasia in a variety of tissues.[432]

Initial progress with PTEN was prompted by the observation that it was frequently mutated in tumours that include glial tumours, breast and prostate cancer.[424,425] Loss of PTEN occurs in 20% of primary tumours and its absence correlated with grade.[433] Loss of heterozygosity at chromosome 10q is observed in 70%–80% of glioblastomas and 20%–40% of glioblastoma multiforme — the most aggressive form of astrocytic tumour.[434,435] In contrast, mutation of the second PTEN allele rarely occurs in lower grade tumours.[436] PTEN mutations are detected at a high frequency in metastasising prostate cancers (>60%) and endometrial cancers most frequently in early stage cases (30%–50%).[437] In tumours inactivating frameshifts, missense mutations and homozygous deletions have been documented leading to premature protein translation. Many of the missense mutations mapping to codons within the catalytic region result in lost phosphatase activity.

11.1.2. *PTEN structure*

PTEN is expressed in most species as a 40–50 kDa protein containing an N-terminal catalytic domain that confers phosphatase activity and a C-terminal phospholipid binding C2 domain (Fig. 9). One exception is found in *C. elegans*, where the PTEN orthologue DAF18 is larger at 120 kDa. PTEN also contains additional motifs that may act as regulatory elements and aid protein translocation. In several species, the N-terminus contains a putative PtdIns(4,5)P_2 binding motif (Lys/Arg-X4-Lys/Arg-X-Lys/Arg-Lys/Arg) which is present in proteins such as villin and gelsolin that regulate the actin cytoskeleton or the plasma membrane. Recent data obtained in *Dictyostelium* indicate that the PtdIns(4,5)P_2-binding motif plays a more important role in membrane localisation of PTEN then originally realised and serves to regulate catalytic activity.[438–440] The crystal structure of PTEN has now been solved and has provided important insight into the regulation of catalytic activity.[441] It reveals that the N-terminal sequence is likely to be exposed and orientated such that it could interact with the membrane bilayer. Mutations within the CRB3 loop of the C domain reduces binding of PTEN to phospholipid vesicles.[441] Furthermore, these mutations suppress anchorage independent growth and proliferation,[442] yet the enzyme retains catalytic activity indicating that the C2 domain independently fulfils an important regulatory role. In higher eukaryotes, the C-terminus also contains a PDZ-binding motif (Thr/Ser-X-Val-COOH) and two PEST motifs which are both dispensable for tumour suppressor function. PDZ-binding motifs are found in several proteins involved in signal transduction, frequently transmembrane receptors or ion channels. These interactions can be of high affinity. PDZ domains can also heterodimerise with PDZ domains of different proteins.[443] The PDZ domain of PTEN mediates its binding to hDLG, hMAST205[444] and MAGI 2 and MAGI 3, two membrane-associated guanylate kinases.[445,446] Phosphorylation of PTEN at five residues of the C-terminus, possibly by CK2,[447–449] alters its binding to PDZ domain-containing proteins indicating a phosphorylation-induced conformational change with a resultant translocation from membrane to cytosol, respectively and increasing its degradation.[450,451] Although the regulatory mechanism remains unknown, PTEN may dephosphorylate these residues.[452]

In addition to protein-protein interactions, PDZ domains in proteins such as syntenin, CASK, Tiam1 and FAP are capable of binding PtdIns(4,5)P_2.[453] In this way, this interaction may also regulate association of PTEN with the plasma membrane. Deletion of the PDZ domain suggested a role for the PDZ domain in platelet-derived growth factor (PDGF)-induced membrane ruffling and cell viability [454] and cell spreading.[455]

11.1.3. *Catalytic activity of PTEN*

Both sporadic and germ line mutations within the putative catalytic domain demonstrated that the domain was required for biological function. Constructs with mutations in the catalytic domain are unable to restore inhibition of cell proliferation to PTEN-deficient cell lines.[431,456,457] Despite this data, significant catalytic activity proved difficult to demonstrate *in vitro*.[458,459] Although PTEN dephosphorylated serine/threonine phosphorylated proteins, its specific activity *in vitro* was very low and poly(Glu-pTyr) proved the most effective substrate. Critically PTEN with inactivating mutations in the catalytic domain retained full activity against poly(Glu-pTyr).[459] On the basis of these data, it was considered that the optimal substrate remained unknown.

11.1.4. *PTEN as a protein phosphatase*

The observation that PTEN decreased phosphorylation of focal adhesion kinase (FAK) both *in vitro* and *in vivo* suggested a mechanism by which PTEN alters the actin cytoskeleton, cell adhesion and motility.[460] Expression of recombinant PTEN inhibited whilst antisense oligonucleotides stimulated fibroblast migration. Binding of PTEN and FAK was demonstrated and found to be dependent on phosphotyrosine residue 397 on FAK, and this interaction regulated the activity of this protein kinase.[461] Similarly, PTEN can dephosphorylate the phosphotyrosine-containing adaptor protein Shc following its immunoprecipitation from cell lysates prepared from cells stimulated with epidermal growth factor.[462] In addition, PTEN binds tyrosine phosphorylated Shc. However, PTEN is a relatively poor protein phosphatase and not all mutations in the catalytic domain abolish its protein phosphate activity.[431,463,464] The observed preference of PTEN protein phosphatase activity towards highly acidic peptides suggested specificity to negatively charged substrates.

11.1.5. *PTEN as a phosphoinositide phosphatase*

The search for alternate substrates continued until it was demonstrated that a subset of inositol phospholipids rather than phosphoproteins were the principle target of PTEN. PTEN dephosphorylates the 3′ phosphate of PtdIns(3,4,5)P_3.[465] Several independent PTEN mutant enzymes preparations from Cowden disease patients were unable to dephosphorylate phosphotyrosine substrates yet retained 3-phosphoinositide phosphatase activity.[463,464] Immortalised cell lines from PTEN knockout mice and PTEN-deficient tumour cell lines have increased levels of intracellular 3-phosphoinositides.[466,467] Given its effect on 3-phosphoinositides, PTEN serves as an endogenous regulator of PI3K signalling pathway. Although PTEN can also utilise other 3-phosphoinositides, the specific activity of PTEN towards these substrates is 50–1500 fold lower than for PtdIns(3,4,5)P3, indicating that it is the preferred substrate *in vivo*. Point mutation of two conserved Lys residues (Lys 125 and Lys 128 in human PTEN) present only in PTEN orthologues demonstrated that they confer its unique substrate specificity.[441] The crystal structure suggests that Lys 125 and Lys 128 interact with the 1′and 5′ phosphates on the inositol ring, respectively.

Over-expression of catalytically active PTEN dramatically inhibited growth factor-induced stimulation of PtdIns(3,4,5)P3 production. In contrast, expression of a kinase inactive PTEN mutant (C124S) increased PtdIns(3,4,5)P3 levels in the absence of receptor stimulation.[463] As such, PTEN serves as an endogenous regulator of the PI3K signalling pathway. Consistent with this role, PTEN expression regulates longevity and dauer formation in *C. elegans*.[468] Expression of the PTEN mutant G129E which inhibits lipid but not protein phosphatase activity does not affect Akt activity.[469]

In addition to its anti-apoptotic effects, PTEN also inhibits cell proliferation. Expression of PTEN results in cell cycle arrest at G1.[469] This effect is mediated via the cyclin-dependent kinase (CDK) inhibitor $p27^{Kip1}$ which inhibits the activity of the CDK2/cyclinE complex.[457] Since CDK2/cyclinE is required for entry into S-phase, then up-regulation of p27kip1 leads to arrest in G1. PTEN also decreases phosphorylation of the retinoblastoma protein (pRb) which is a substrate of CDK2 suggesting a role of pRb in PTEN-induced cell cycle arrest.[470]

11.1.6. *Regulation of PTEN catalytic activity*

Phosphorylation of three residues serine 380, threonine 382 and threonine 383 at the C-terminus is required to maintain enzyme stability and attenuate catalytic activity.[449,450,471,472] This phosphorylation is thought to flip the PTEN structure to a closed monomeric configuration. Dephosphorylation leads to the recruitment of PTEN into high molecular mass complexes containing scaffold proteins such as MAGI-2 considered to be important for localisation at the plasma membrane.[446,473]

11.2. Myotubularin phosphatase family

Mutations in the MTM1 gene located on chromosome Xq28 are responsible for X-linked myotubular myopathy (MTM). MTM is a degenerative disease of muscle characterised by lack of muscle tone and weakness.[474,475] There is a wide spectrum of symptoms in affected individuals, but when severe can be fatal due to respiratory failure. Histopathological analysis reveals the underlying cause is arrested myotube differentiation with muscle of patients showing a pattern of protein expression that includes foetal myosin, desmin, vimentin and nerve cell adhesion molecule normally found in foetal myotubes.[476,477] In those patients, a large number of mutations affect the catalytic domain of the MTM1 gene product where truncation of the catalytic domain correlates with severe disease phenotype. MTM1 is one member of a large gene family encoding 14 myotubularin-like proteins.[478] Eight isoforms are catalytically active with six containing inactivating mutations within the catalytic site.[479] Mutations in the *MTMR2* gene is responsible for a second neuromuscular disorder Type 4B Charcot-Marie Tooth syndrome characterised by defects in myelination of neurones and the proliferation of Schwann cells in peripheral nerves.[480,481] Myotubularin-related 5 (MTMR5) also known as Set binding factor 1 is a non-catalytic myotubularin but its knockout in mice renders them infertile.[482]

11.2.1. *Myotubularin structure*

The catalytic domain of myotubularin-related proteins have sequence homology with dual specificity phosphatases and are also active on phosphotyrosine, phosphoserine and phosphothreonine residues[483] (Fig. 9). Each myotubularin has an N-terminal PH domain within a region that was originally described as a GRAM (Glucose transferase, Rab like GTPase activator and Myotubularin) domain.[484,485] Consequently it is also referred to as the PH-GRAM domain. The catalytic domain of myotubularins is larger than that found in other protein tyrosine phosphatases (375 amino acid residues). Within the catalytic domain, additional motifs have been described. These include a rac interacting domain (RID) and a SET interaction domain (SID). C-terminal to the catalytic domain is a coiled-coil domain that mediates heterodimerisation of myotubularins.[486,487] At that C-terminus, many tubularins contain a phosphoinositide- or phospholipid-binding domain that include the PDZ domain, a FYVE domain or a PH domain. However, in the case of MTMR3 it is the PH-GRAM domain that localises this myotubularin at the membrane[488] where it binds PtdIns(5)P which also serves to increase specific activity. Of interest is the observation that localisation of MTM1 to late endosomes is responsive to EGF stimulation.[489]

11.2.2. *Regulation of MTM-related phosphatase activity*

In a way that parallels PTEN, MTM1 was characterised in detail because it was a poor phosphoprotein phosphatase. Sequence similarity of MTM1 and the yeast phosphoinositide phosphatase Sac1p provided more evidence that MTM1 may have an alternative substrate specificity. Consequently, PtdIns(3)P was found to serve as the best substrate for recombinant MTM1.[490] Expression of recombinant kinase inactive MTM1 increases basal levels of PtdIns(3)P. Phosphatase activity against PtdIns(3)P has now been extended to the other MTM1-related enzymes. In contrast to PTEN, the substrate specificity of MTM1-related phosphatases is severely restricted.

11.3. SH2 domain containing inositol 5′phosphatase (SHIP)

The SH2 domain containing inositol 5′phosphatase (SHIP) was originally isolated as a 145 kDa protein that was tyrosine phosphorylated and associated with the adaptor protein Shc following stimulation of haematopoietic cells with growth factors, cytokines and immunoglobulins.[491] SHIP is a phosphatase that utilises PtdIns(3,4,5)P_3 and inositol 1,3,4,5 tetrakisphosphate (IP4) *in vitro* and is thought to produce PtdIns(3,4)P_2 *in vivo*.[492] In addition, there are two additional SHIP enzymes. sSHIP is a 104 kDa protein expressed in haematopoietic and embryonic stem cells.[493] sSHIP is generated by transcription from a promoter within the intron between exons 5 and 6 of the SHIP1 gene. In contrast, SHIP2 is encoded by a distinct gene and it ubiquitously expresses a 150 kDa enzyme.[494] All SHIP enzymes are alternatively spliced and C-terminal truncations have been reported, however their biological relevance remains unclear.[495,496]

11.3.1. *SHIP structure*

SHIP and SHIP2 contain an N-terminal SH2 domain that binds the consensus sequence pY(Y/D)X(L/I/V)[497] (Fig. 9). Downstream from this is the 5-phosphatase domain followed by two NPXY sequences that are phosphorylated and act as binding motifs for phosphotyrosine binding (PTB) domains of Shc and Dok,[498,499] and a C-terminus that is rich in proline residues that can bind SH3 domain-containing proteins such as Grb2 and the tyrosine kinases Src and Abl.[500]

11.3.2. *Regulation of SHIP phosphatase activity*

Since the phosphatase activity of neither SHIP nor SHIP2 alter following stimulation with growth factors or tyrosine phosphorylation it is thought that the translocation of SHIP to sites of PtdIns(3,4,5)P_3 generation regulate its inhibitory effects.[501] The SH2 domain of SHIP binds pY239 and pY317 of Shc[502] whilst the PTB domain of Shc bind the NPXpY motifs of SHIP.[503,504] The recruitment of SHIP to scaffold proteins such as Gab and Dok results in the presence of these phosphatases in large multimeric complexes.[505,506] Both the SH2 domain and polyproline rich C-terminus of SHIP are required for the translocation of SHIP to the plasma membrane and biological activity.[507] Shc has been proposed to mediate the recruitment of SHIP to the plasma membrane.[508] Here SHIP binds receptors bearing either immunoreceptor tyrosine-based inhibitory motifs (ITIM) such as FcγRIIB[509] and receptors with immunoreceptor tyrosine-based activation motifs (ITAMs) such as the ζ chain of the T-cell receptor, the β and γ subunits of the FcεRI[497,510] or FcγRIIa.[511] In this way together with PTEN, SHIP phosphatase activity serves to attenuate receptor-mediated PI3K activation (Fig. 11).

11.3.3. *SHIP knockout mice*

SHIP null mice have elevated levels of PtdIns(3,4,5)P_3but are viable although they do have a shortened lifespan.[512] These mice have elevated numbers of granulocytes and macrophages, suffer from progressive splenomegaly, extramedullary homopoiesis, myeloid infiltration of the lungs and osteoporosis due to increased numbers of osteoclasts.[512,513] Granulocyte-macrophage progenitor cells are expanded and become hypersensitive to growth factors. Although SHIP−/− mice share several phenotypic similarities with PTEN+/− mice consistent due to elevated PtdIns(3,4,5)P_3 levels generation of PTEN+/− SHIP−/− mice developed a myelodysplasia-like syndrome, characterised by marked leukocytosis, anaemia and thrombocytopenia not observed in SHIP knockout mice.[514]

Mice deficient in SHIP2 have a more severe phenotype than SHIP−/− mice and die from insulin hypersensitivity-induced hypoglycaemia.[515] Mutations in the SHIP2 gene are associated with type II diabetes.[516] No difference was observed in the coding region but one patient had a 16 bp deletion in the proximal part of the SHIP2 3′untranslated region, this results

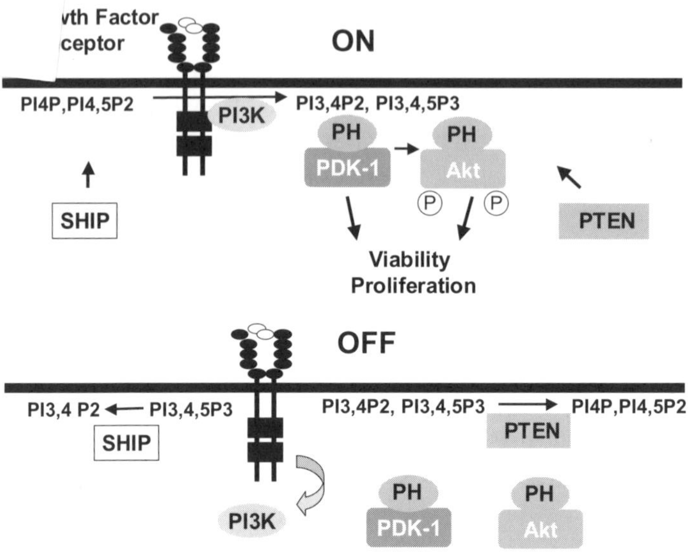

Fig. 11 PTEN- and SHIP-regulated receptor-mediated 3-phosphoinositide production. Stimulation of receptors at the plasma membrane results in the recruitment of PI3K enzymes and the generation of PtdIns(3,4,5)P_3. This serves to translocate both PDK-1 and Akt kinases via their PH domains and serves to promote cell viability and proliferation by phosphorylation of their downstream targets. Recruitment of SHIP and PTEN to the signalling complex serves to hydrolyse PtdIns(3,4,5)P_3 to PtdIns(3,4)P_2 and both PtdIns(3,4,5)P_3 and PtdIns(3,4)P_2 to PtdIns(4,5)P_2 and PtdIns(4)P, respectively. This attenuates the localisation of PDK-1 and Akt at the membrane and their kinase activity. The recruitment of protein tyrosine phosphatases such as SHP dephosphorylates tyrosine residues leading to the dissociation of the signalling complex.

in increased expression of SHIP2 and decreased insulin sensitivity. Together PTEN, SHIP and myotubularin phosphatases regulate the cellular content of 3-phosphoinositides produced by PI3K activity and their action is critical in maintaining a normal phenotype (Fig. 12).

12. The Role of the PI3K Pathway in Disease

12.1. Cancer

Constitutive activation of tyrosine kinase activity in many human tumours results in deregulated stimulation of the PI3K pathway. Equivalent

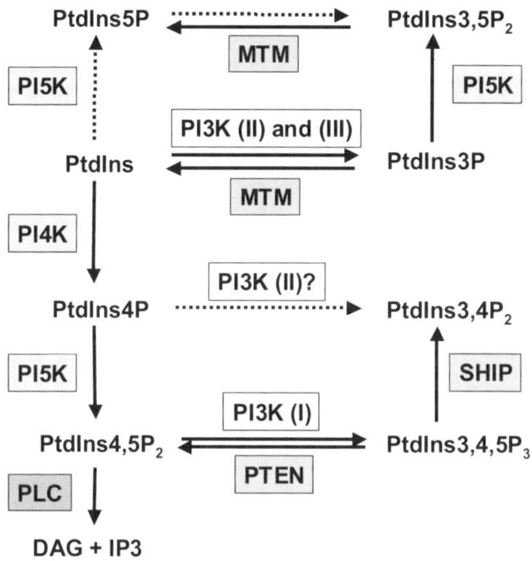

Fig. 12 The role of phosphoinositide phosphatases in 3-phosphoinositide production. The intimate relationship between phosphoinositide kinases and phosphoinositide phosphatases PTEN, SHIP and SHIP2 and the myotubularin phosphatase family (MTM) is shown. Since PtdIns(3,4,5)P_3 serves as a substrate for both SHIP and PTEN, their action serves to attenuate signalling downstream of class I and possibly class II PI3K enzymes. Hydrolysis of PtdIns(3)P and PtdIns(3,5)P_2 by MTM phosphatases limits signalling pathways initiated by the action of class II and class III PI3K enzymes.

phenotypes are achieved by activating mutations of the catalytic and regulatory subunits or gene amplification.

The first evidence of a direct role of PI3K enzymes in cancer was provided by the identification of the retroviral oncogene product v-p3k, which encodes the catalytic subunit of p110α responsible for haemangiosarcomas in birds.[517] However, its cellular counterpart, c-P3k, lacks oncogenicity. Fusion of a viral Gag sequence to the amino terminus of c-P3k activates the transforming potential. Activation was also obtained by the addition of other membrane localisation signals. Furthermore, mutations that inactivate lipid kinase activity abolish oncogenicity. The transforming activity of P3k is correlated with the ability to induce activating phosphorylation in Akt. Interactions of P3k with the regulatory subunit of PI3K, p85, or with Ras are not required for transformation. In this way, oncogenicity of P3k depends on constitutive lipid kinase activity.[160,518] The gene PIK3A that encodes p110α maps to chromosome 3q26, a locus amplified in many human cancers. Specifically, somatic mutation and/or gain of PIK3CA copy number have been described in many cancers of the head and neck, colon, stomach, lung, ovary, cervix, breast, thyroid and glioblastomas.[519–523] The majority of mutations in PIK3A ($>80\%$) map to three discrete clusters — the p85 binding region (exon 1), the PIK domain (exon 9) and the kinase domain (exon 20).[522] A high frequency of PIK3CA mutation was reported in many cancers, including colon (32%), brain (27%), stomach (25%), lung (4%) and breast (20%).[522,524] The increase in copy number of the PIK3CA gene is associated with increased transcription and expression of the p110α enzyme.[520,525]

Cloning of p65-PI3K, identified a mutant of the class IA PI3K regulatory subunit which includes the initial 571 residues of the wild type p85α-protein linked to a region conserved in the eph tyrosine kinase receptor family.[526] This mutation, obtained from a transformed cell, induced constitutive activation of PI3K and contributed to cellular transformation. Transgenic mice expressing p65 in T lymphocytes developed an infiltrating lymphoproliferative disorder and autoimmune renal disease with an increased number of T lymphocytes exhibiting a memory phenotype and reduced apoptosis.[527] This pathology was strikingly similar to that described in mice exhibiting heterozygous loss of the tumor suppressor PTEN, a lipid and protein phosphatase. Immunoprecipitation of p85α with polyclonal antiserum revealed increased lipid kinase activity in 86% of colorectal tumours.[528] Somatic mutations in the class IA PI3K adaptor p85α together with constitutive activation of the catalytic subunit have been reported in tumours of the colon and ovary.[529] The role of PtdIns(3,4,5)P_3 in cell migration has important implications for tumour metastasis, and the PI3K-Akt signalling pathway has been associated with the metastasis of several tumours.[530] Both the Rac GTPase and ILK have been proposed as possible mediators of increased tumour invasion downstream of $\alpha6\beta4$ integrin-stimulated PtdIns(3,4,5)P_3 production.[531]

Beclin1 is the human homologue of Apg6, a vps34p-binding protein is deleted in 40%–75% of breast cancers.[532] Consequently, it is possible that Beclin1-mediated tumour suppression is dependent upon vps34p/vps15p-mediated autophagy, Interestingly, characterisation of mTOR as an important downstream regulator of the PI3K signalling pathway is supported by the observation that cells derived from both mammalian and avian tumours either lacking functional PTEN or activated forms of Akt or PI3K are sensitive to rapamycin or its analogs.[533,534] These data indicate that these tumours are dependent on mTOR which contributes to their proliferation or cell viability.

The role of Akt in cancer was clear from its original description as a retroviral oncogene whose gene product v-Akt contains an N-terminal Gag sequence that facilitates membrane localisation and confers oncogenicity.[535] Its activation by PtdIns(3,4,5)P_3 means that Akt mediates tumourigenesis in those cancers that have deregulated PI3K activity or a loss of PTEN. Amplification of the Akt1 gene has been reported in glioblastoma,[536] Akt1 and Akt2 genes are amplified in gastric adenocarcinoma[537] and the Akt2 gene is amplified in ovarian, pancreatic, gastric and breast cancer.[538,539] Increased Akt2 activity was reported in 40% of primary ovarian cancers.[540] Analysis of Akt3 gene expression suggests a possible correlation with advanced breast and prostate cancer.[541] Constitutively active Akt has been detected in several cancers including breast, colon, pancreatic, prostate and bile duct cancer.[542,543] Active Akt expression correlates with poor prognosis for many cancers.[544–547] The relevance of the tumour suppressor activity of PTEN and its relevance to PI3K signalling and cancer have already been described, suffice to note that expression of recombinant PTEN induces apoptosis of breast cancer, prostate cancer and leukaemia cell lines.[431,463,548–552] Fibroblasts derived from PTEN null mice are highly resistant to apoptotic stimuli.[466]

12.2. Autoimmuninty

Mice expressing a p85 deletion mutant termed p65 contain constitutively active p110 in T-cells.[527] These animals have expansion of CD4+ T-cell numbers in lymph nodes and spleen together with high numbers of memory cell markers. Although expression of p65 was driven by the T-cell specific Lck promoter, these mice have hypergammaglobulinaemia with increased titres of IGG1, IgG2a and anti-dsDNA autoantibody titres. There are immune deposits in the kidney and lymphoid infiltrates in many tissues but there was no evidence of increased cell death or thymocyte deletion.[553] In contrast, p65 expression decreased CD4-Cd8− T-cell numbers and increased the CD4+Cd8+ T-cell population.

Mice lacking p110γ showed the opposite phenotype with B-cells unaffected. Mice null for p85 or p110δ reveal defects in marginal zone B-cells and development of CD5+B1 B-cells.[334,356,365,554–556] Mice null for p85 show a phenotype similar to Xid, the murine equivalent of human XLA. Mice lacking p110δ also have defects in T-cell antigen receptor signalling.[364,557] Mice heterozygous for PTEN develop a wide range of tumours in addition to autoimmune disease with lymphadenopathy and splenomegaly.[558]

13. Concluding Remarks

In the past 17 years, since the first description of phosphatidylinositol-3-phosphate,[28] there have been thousands of publications that describe intracellular 3-phosphoinositides, their roles as intracellular signalling molecules, the enzymes responsible for their production and downstream targets. Despite this intense scrutiny, many challenges still remain. The role of the PI3K signalling pathway in numerous pathologies remains clear. We still await improved PI3K isozyme selective small molecule inhibitors that may allow therapeutic intervention in cancer and autoimmune disease. The full implication of the class II and class III PI3K isoforms awaits clarification. Issues regarding the spatial-temporal production of 3-phosphoinositides require elucidation. Doubtless, recent developments in the chemistry of fluorescent proteins and dyes will help progress towards this goal. The analysis of class I PI3K knockout mice proceeds apace yet mice null for class II and class III PI3K have yet to be described. Today, a localised production of PtdIns(3)P is being considered in the context of a receptor-mediated signalling response and so the full potential of the other 3-phosphoinositides may become expanded. In recent years, the description of 3-phosphoinositide phosphatases has revolutionised the way we consider PI3K signalling in cancer and autoimmune disease. The full implication of these phosphates in physiology and disease still requires further investigation together with the development of specific antagonists for both research and therapeutic use. Consequently, although work on the PI3K signalling pathway has almost come of age, there remain many questions to answer before we even begin to consider early retirement.

References

1. M.R. Hokin and L.E. Hokin, Enzyme secretion and the incorporation of P32 into phospholipides of pancreas slices, *J. Biol. Chem.* **203** (1953) 967–977.
2. L.E. Hokin and M. Hokin-Neaverson, Commentary on 'Effects of acetylcholine on the turnover of phosphoryl units in individual phospholipids of pancreas slices and brain cortex slices', *Biochim. Biophys. Acta* **1000** (1989) 465–469.
3. L.E. Hokin, Dynamic aspects of phospholipids during protein secretion, *Int. Rev. Cytol.* **23** (1968) 187–208.
4. J. Durell, J.T. Garland and R.O. Friedel, Acetylcholine action: biochemical aspects, *Science* **165** (1969) 862–866.
5. E.G. Lapetina and R.H. Michell, Phosphatidylinositol metabolism in cells receiving extracellular stimulation, *FEBS Lett.* **31** (1973) 1–10.
6. R.H. Michell, S.S. Jafferji and L.M. Jones, Receptor occupancy dose — response curve suggests that phosphatidyl-inositol breakdown may be intrinsic to the mechanism of the muscarinic cholinergic receptor, *FEBS Lett.* **69** (1976) 1–5.
7. R.H. Michell, Inositol phospholipids and cell surface receptor function, *Biochim. Biophys. Acta* **415** (1975) 81–147.
8. R.A. Akhtar and A. Abdel-Latif, Requirement for calcium ions in acetylcholine-stimulated phosphodiesteratic cleavage of phosphatidyl-myo-inositol 4,5-bisphosphate in rabbit iris smooth muscle, *Biochem. J.* **192** (1980) 783–791.
9. M.J. Berridge, Rapid accumulation of inositol trisphosphate reveals that agonists hydrolyse polyphosphoinositides instead of phosphatidylinositol, *Biochem. J.* **212** (1983) 849–858.
10. M.J. Berridge, R.M. Dawson, C.P. Downes, J.P. Heslop and R.F. Irvine, Changes in the levels of inositol phosphates after agonist-dependent hydrolysis of membrane phosphoinositides, *Biochem. J.* **212** (1983) 473–482.
11. H. Streb, R.F. Irvine, M.J. Berridge and I. Schulz, Release of Ca^{2+} from a nonmitochondrial intracellular store in pancreatic acinar cells by inositol-1,4,5-trisphosphate, *Nature* **306** (1983) 67–69.
12. H. Stenmark and D.J. Gillooly, Intracellular trafficking and turnover of phosphatidylinositol 3-phosphate, *Semin. Cell Dev. Biol.* **12** (2001) 193–199.
13. L.R. Stephens, T.R. Jackson and P.T. Hawkins, Agonist-stimulated synthesis of phosphatidylinositol(3,4,5)-trisphosphate: a new intracellular signalling system? *Biochim. Biophys. Acta* **1179** (1993) 27–75.

14. S.K. Dove, F.T. Cooke, M.R. Douglas, L.G. Sayers, P.J. Parker and R.H. Michell, Osmotic stress activates phosphatidylinositol-3,5-bisphosphate synthesis 5, *Nature* **390** (1997) 187–192.
15. J.J. Hsuan, S. Minogue and S.M. Dos, Phosphoinositide 4- and 5-kinases and the cellular roles of phosphatidylinositol 4,5-bisphosphate, *Adv. Cancer Res.* **74** (1998) 167–216.
16. P.A. Oude Weernink, M. Schmidt and K.H. Jakobs, Regulation and cellular roles of phosphoinositide 5-kinases, *Eur. J. Pharmacol.* **500** (2004) 87–99.
17. T. Gehrmann and L.M. Heilmeyer, Jr., Phosphatidylinositol 4-kinases, *Eur. J. Biochem.* **253** (1998) 357–370.
18. Y. Nishizuka, Turnover of inositol phospholipids and signal transduction, *Science* **225** (1984) 1365–1370.
19. Y. Nishizuka, The role of protein kinase C in cell surface signal transduction and tumour promotion, *Nature* **308** (1984) 693–698.
20. Y. Sugimoto, M. Whitman, L.C. Cantley and R.L. Erikson, Evidence that the Rous sarcoma virus transforming gene product phosphorylates phosphatidylinositol and diacylglycerol, *Proc. Natl. Acad. Sci. USA* **81** (1984) 2117–2121.
21. I.G. Macara, G.V. Marinetti and P.C. Balduzzi, Transforming protein of avian sarcoma virus UR2 is associated with phosphatidylinositol kinase activity: possible role in tumorigenesis, *Proc. Natl. Acad. Sci. USA* **84** (1984) 2728–2732.
22. K.R. Auger, L.A. Serunian, S.P. Soltoff, P. Libby and L.C. Cantley, PDGF-dependent tyrosine phosphorylation stimulates production of novel polyphosphoinositides in intact cells, *Cell* **57** (1989) 167–175.
23. L. Varticovski, B. Druker, D. Morrison, L. Cantley and T. Roberts, The colony stimulating factor-1 receptor associates with and activates phosphatidylinositol-3 kinase, *Nature* **342** (1989) 699–702.
24. M. Whitman, D.R. Kaplan, B. Schaffhausen, L. Cantley and T.M. Roberts, Association of phosphatidylinositol kinase activity with polyoma middle-T competent for transformation, *Nature* **315** (1985) 239–242.
25. S.A. Courtneidge and A. Heber, An 81 kd protein complexed with middle T antigen and pp60c-src: a possible phosphatidylinositol kinase, *Cell* **50** (1987) 1031–1037.
26. D.A. Talmage, R. Freund, A.T. Young, J. Dahl, C.J. Dawe and T.L. Benjamin, Phosphorylation of middle T by pp60c-src: a switch for binding of phosphatidylinositol 3-kinase and optimal tumorigenesis, *Cell* **59** (1989) 55–65.
27. G. Endemann, S.N. Dunn and L.C. Cantley, Bovine brain contains two types of phosphatidylinositol kinase, *Biochemistry* **26** (1987) 6845–6852.

28. M. Whitman, C.P. Downes, M. Keeler, T. Keller and L. Cantley, Type I phosphatidylinositol kinase makes a novel inositol phospholipid, phosphatidylinositol-3-phosphate, *Nature* **332** (1988) 644–646.
29. A. Arcaro and M.P. Wymann, Wortmannin is a potent phosphatidylinositol 3-kinase inhibitor: the role of phosphatidylinositol 3,4,5-trisphosphate in neutrophil responses, *Biochem. J.* **296** (1993) 297–301.
30. C.J. Vlahos, W.F. Matter, K.Y. Hui and R.F. Brown, A specific inhibitor of phosphatidylinositol 3-kinase, 2-(4-morpholinyl)-8-phenyl-4H-1-benzopyran-4-one (LY294002), *J. Biol. Chem.* **269** (1994) 5241–5248.
31. M.J. Cross, A. Stewart, M.N. Hodgkin, D.J. Kerr and M.J. Wakelam, Wortmannin and its structural analogue demethoxyviridin inhibit stimulated phospholipase A2 activity in Swiss 3T3 cells. Wortmannin is not a specific inhibitor of phosphatidylinositol 3-kinase, *J. Biol. Chem.* **270** (1995) 25352–25355.
32. M.P. Wymann, G. Bulgarelli-Leva, M.J. Zvelebil, L. Pirola, B. Vanhaesebroeck, M.D. Waterfield and G. Panayotou, Wortmannin inactivates phosphoinositide 3-kinase by covalentmodification of Lys-802, a residue involved in the phosphate transfer reaction, *Mol. Cell. Biol.* **16** (1996) 1722–1733.
33. R. Yao and G.M. Cooper, Requirement for phosphatidylinositol-3 kinase in the prevention of apoptosis by nerve growth factor, *Science* **267** (1995) 2003–2006.
34. R.M. Schultz, R.L. Merriman, S.L. Andis, R. Bonjouklian, G.B. Grindey, P.G. Rutherford, A. Gallegos, K. Massey and G. Powis, In vitro and in vivo antitumor activity of the phosphatidylinositol-3-kinase inhibitor, wortmannin, *Anticancer Res.* **15** (1995) 1135–1139.
35. L.E. Lemke, G.D. Paine-Murrieta, C.W. Taylor and G. Powis, Wortmannin inhibits the growth of mammary tumors despite the existence of a novel wortmannin-insensitive phosphatidylinositol-3-kinase, *Cancer Chemother. Pharmacol.* **44** (1999) 491–497.
36. S. Semba, N. Itoh, M. Ito, M. Harada and M. Yamakawa, The in vitro and in vivo effects of 2-(4-morpholinyl)-8-phenyl-chromone (LY294002), a specific inhibitor of phosphatidylinositol 3′-kinase, in human colon cancer cells, *Clin. Cancer Res.* **8** (2002) 1957–1963.
37. L. Hu, C. Zaloudek, G.B. Mills, J. Gray and R.B. Jaffe, In vivo and in vitro ovarian carcinoma growth inhibition by a phosphatidylinositol 3-kinase inhibitor (LY294002), *Clin. Cancer Res.* **6** (2000) 880–886.

38. K. Yu, J. Lucas, T. Zhu, A. Zask, C. Gaydos, L. Toral-Barza, J. Gu, F. Li, I. Chaudhary, P. Cai, J. Lotvin, R. Petersen, M. Ruppen, M. Fawzi, S. yral-Kaloustian, J. Skotnicki, T. Mansour, P. Frost and J. Gibbons, PWT-458, A novel pegylated-17-hydroxywortmannin, inhibits phosphatidylinositol 3-kinase signaling and suppresses growth of solid tumors, *Cancer Biol. Ther.* **4** (2005) 538–545.
39. L.C. Foukas, N. Daniele, C. Ktori, K.E. Anderson, J. Jensen and P.R. Shepherd, Direct effects of caffeine and theophylline on p110 delta and other phosphoinositide 3-kinases. Differential effects on lipid kinase and protein kinase activities, *J Biol. Chem.* **277** (2002) 37124–37130.
40. S.J. Morgan, A.D. Smith and P.J. Parker, Purification and characterization of bovine brain type I phosphatidylinositol kinase, *Eur. J. Biochem.* **191** (1990) 761–767.
41. M. Otsu, I.D. Hiles, I. Gout, M.J. Fry, F. Ruiz-Larrea, G. Panayotou, A. Thompson, R. Dhand, J. Hsuan, N.F. Totty, A.D. Smith, S.J. Morgan, S.A. Courtneidge, P.J. Parker and M.D. Waterfield, Chracterization of two 85 kd proteins that associate with receptor tyrosine kinases, middle-T/pp60c-src complexes, and PI3Kinase, *Cell* **65** (1991) 91–104.
42. E.Y. Skolnik, B. Margolis, M. Mohammadi, E. Lowenstein, R. Fisher, A. Drepps, A. Ullrich and J. Schlessinger, Cloning of PI3Kinase-associated p85 utilizing a novel method for expression/cloning of target proteins for receptor tyrosine kinases, *Cell* **65** (1991) 83–90.
43. J. Escobedo, S. Navankasattusas, W.M. Kavanaugh, D. Milfay, V.A. Fried and L.T. Williams, cDNA cloning of a noval 85 kd protein that has SH2 domains and regulates binding of PI3Kinase to the PDGF beta-receptor, *Cell* **65** (1991) 75–82.
44. I.D. Hiles, M. Otsu, S. Volinia, M.J. Fry, I. Gout, R. Dhand, G. Panayotou, F. Ruiz-Larrea, A. Thompson, N.F. Totty, J. Justin Hsuan, S.A. Courtneidge, P.J. Parker and M.D. Waterfield, Phosphatidylinositol 3-kinase: structure and expression of the 110 kd catalytic subunit, *Cell* **70** (1992) 419–429.
45. L.R. Stephens, A. Eguinoa, H. Erdjument-Bromage, M. Lui, F. Cooke, J. Coadwell, A.S. Smrcka, M. Thelen, K. Cadwallader, P. Tempst and P.T. Hawkins, The G beta gamma sensitivity of a PI3K is dependent upon a tightly associated adaptor, p101, *Cell* **89** (1997) 105–114.
46. P.V. Schu, K. Takegawa, M.J. Fry, J.H. Stack, M.D. Waterfield and S.D. Emr, Phosphatidylinositol 3-kinase encoded by the yeast VPS34 gene essential for protein sorting, *Science* **260** (1993) 88–92.

47. A.L. Munn and H. Reizman, Endocytosis is required for the growth of vacuolar H^+-ATPase-defective yeast: identification of six new END genes, *J. Cell Biol.* **127** (1994) 373–385.
48. P. Hu, A. Mondino, E.Y. Skolnik and J. Schlessinger, Cloning of a novel, ubiquitously expressed human phosphatidylinositol 3-kinase and identification of its binding site on p85, *Mol. Cell. Biol.* **13** (1993) 7677–7688.
49. B. Vanhaesebroeck, M.J. Welham, K. Kotani, R. Stein, P.H. Warne, M.J. Zvelebil, K. Higashi, S. Volinia, J. Downward and M.D. Waterfield, p110δ, a novel phosphoinositide 3-kinase in leukocytes, *Proc. Natl. Acad. Sci. USA* **94** (1997) 4330–4335.
50. L.K. MacDougall, J. Domin and M.D. Waterfield, A family of phosphoinositide 3-kinases in Drosophila identifies a new mediator of signal transduction, *Curr. Biol.* **5** (1995) 1404–1415.
51. J. Domin, F. Pages, S. Volinia, S.E. Rittenhouse, M.J. Zvelebil, R.C. Stein and M.D. Waterfield, Cloning of a human phosphatidylinositol 3-kinase with a C2 domain which displays reduced sensitivity to the inhibitor wortmannin, *Biochem. J.* **326** (1997) 139–147.
52. R.A. Brown, L.K. Ho, S.J. Weber-Hall, J.M. Shipley and M.J. Fry, Identification and cDNA cloning of a novel mammalian C2 domain-containing phosphoinositide 3-kinase, HsC2-PI3K, *Biochem. Biophys. Res. Commun.* **233** (1997) 537–544.
53. A. Arcaro, S. Volinia, M.J. Zvelebil, R. Stein, S.J. Watton, M.J. Layton, I. Gout, K. Ahmadi, J. Downward and M.D. Waterfield, Human PI3Kinase C2β — the role of calcium and the C2 domain in enzyme activity, *J. Biol. Chem.* **273** (1998) 33082–33091.
54. M. Rozycka, Y.J. Lu, R.A. Brown, M.R. Lau, J.M. Shipley and M.J. Fry, cDNA cloning of a third human C2-domain-containing class II phosphoinositide 3-kinase, PI3K-C2gamma, and chromosomal assignment of this gene (PIK3C2G) to 12p12, *Genomics* **54** (1998) 569–574.
55. J.V. Virbasius, A. Guilherme and M.P. Czech, Mouse p170 is a novel phosphatidylinositol 3-kinase containing a C2 domain, *J. Biol. Chem.* **271** (1996) 13304–13307.
56. L. Molz, Y.W. Chen, M. Hirano and L.T. Williams, Cpk is a novel class of drosophila PtdIns 3-kinase containing a C2 domain, *J. Biol. Chem.* **271** (1996) 13892–13899.
57. F. Ono, T. Nakagawa, S. Saito, Y. Owada, H. Sakagami, K. Goto, M. Suzuki, S. Matsuno and H. Kondo, A novel class II phosphoinositide 3-kinase

predominantly expressed in the liver and its enhanced expression during liver regeneration, *J. Biol. Chem.* **273** (1998) 7731–7736.

58. M.J. Zvelebil, L. MacDougall, S. Leevers, S. Volinia, B. Vanhaesebroeck, I. Gout, G. Panayotou, J. Domin, R. Stein, F. Pages, H. Koga, K. Salim, J. Linacre, P. Das, C. Panaretou, R. Wetzker and M.D. Waterfield, Structural and functional diversity of phosphoinositide 3-kinases, *Philos. Trans. R. Soc. Lond. B* **351** (1996) 217–223.
59. Z. Songyang, S.E. Shoelson, M. Chaudhuri, G. Gish, T. Pawson, W.G. Haser, F. King, T. Roberts, S. Ratnofsky, R.J. Lechleider, B.G. Neel, R.B. Birge, J.E. Fajardo, M.M. Chou, H. Hanafusa, B. Schaffhausen, L.C. Cantley, SH2 domains recognize specific phosphopeptide sequences, *Cell* **72** (1993) 767–778.
60. R. Dhand, K. Hara, I. Hiles, B. Bax, I. Gout, G. Panayotou, M.J. Fry, K. Yonezawa, M. Kasuga and M.D. Waterfield, PI 3-kinase: structural and functional analysis of intersubunit interactions, *EMBO J.* **13** (1994) 511–521.
61. R. Dhand, I. Hiles, G. Panayotou, S. Roche, M.J. Fry, I. Gout, N.F. Totty, O. Truong, P. Vicendo, K. Yonezawa, M. Kasuga, S.A. Courtneidge and M.D. Waterfield, PI 3-kinase is a dual specificity enzyme: autoregulation by an intrinsic protein-serine kinase activity, *EMBO J.* **13** (1994) 522–533.
62. L. Bunday, Membrane-targeting of signalling molecules by SH2/SH3 domain-containing adaptor proteins, *Biochim. Biophys. Acta* **1422** (1999) 187–204.
63. G. Pasternak, A. Hochhaus, B. Schultheis and R. Hehlmann, Chronic myelogenous leukemia: molecular and cellular aspects, *J. Cancer Res. Clin. Oncol.* **124** (1998) 643–660.
64. Y. Zheng, S. Bagrodia and R.A. Cerione, Activation of phosphoinositide 3-kinase activity by Cdc42Hs binding to p85, *J. Biol. Chem.* **269** (1994) 18727–18730.
65. K.F. Tolias, L.C. Cantley and C.L. Carpenter, Rho family GTPases bind to phosphoinositide kinases, *J. Biol. Chem.* **270** (1995) 17656–17659.
66. K.M. Hill, Y. Huang, S.C. Yip, J. Yu, J.E. Segall and J.M. Backer, N-terminal domains of the class ia phosphoinositide 3-kinase regulatory subunit play a role in cytoskeletal but not mitogenic signaling, *J. Biol. Chem.* **276** (2001) 16374–16378.
67. D.A. Antonetti, P. Algenstaedt and C.R. Kahn, Insulin receptor substrate 1 binds two novel splice variants of the regulatory subunit of phosphatidylinositol 3-kinase in muscle and brain, *Mol. Cell. Biol.* **16** (1996) 2195–2203.

68. K. Inukai, M. Anai, E. Van Breda, T. Hosaka, H. Katagiri, M. Funaki, Y. Fukushima, T. Ogihara, Y. Yazaki, Kikuchi, Y. Oka and T. Asano, A novel 55-kDa regulatory subunit for phosphatidylinositol 3-kinase structurally similar to p55PIK Is generated by alternative splicing of the p85alpha gene, *J. Biol. Chem.* **271** (1996) 5317–5320.
69. D.A. Fruman, L.C. Cantley and C.L. Carpenter, Structural organization and alternative splicing of the murine phosphoinositide 3-kinase p85 alpha gene, *Genomics* **37** (1996) 113–121.
70. K. Inukai, M. Funaki, T. Ogihara, H. Katagiri, A. Kanda, M. Anai, Y. Fukushima, T. Hosaka, M. Suzuki, B.C. Shin, K. Takata, Y. Yazaki, M. Kikuchi, Y. Oka and T. Asano, p85alpha gene generates three isoforms of regulatory subunit for phosphatidylinositol 3-kinase (PI 3-Kinase), p50alpha, p55alpha, and p85alpha, with different PI 3-kinase activity elevating responses to insulin, *J. Biol. Chem.* **272** (1997) 7873–7882.
71. S. Pons, T. Asano, E. Glasheen, M. Miralpeix, Y. Zhang, T.L. Fisher, M.G. Myers, X.J. Sun and M.F. White, The structure and function of p55PIK reveal a new regulatory subunit for phosphatidylinositol 3-kinase, *Mol. Cell. Biol.* **15** (1995) 4453–4465.
72. N.J. Kerouz, D. Horsch, S. Pons and C.R. Kahn, Differential regulation of insulin receptor substrates-1 and -2 (IRS-1 and IRS-2) and phosphatidylinositol 3-kinase isoforms in liver and muscle of the obese diabetic (ob/ob) mouse, *J. Clin. Invest.* **100** (1997) 3164–3172.
73. X. Xia and G. Serrero, Multiple forms of p55PIK, a regulatory subunit of phosphoinositide 3 kinase, are generated by alternative initiation of translation, *Biochem. J.* **341**(Pt 3) (1999) 831–837.
74. P. Rodriguez-Viciana, P.H. Warne, R. Dhand, B. Vanhaesebroeck, I. Gout, M.J. Fry, M.D. Waterfield and J. Downward, Phosphatidylinositol-3-OH kinase as a direct target of Ras, *Nature* **370** (1994) 527–532.
75. P. Rodriguez-Viciana, P.H. Warne, B. Vanhaesebroeck, M.D. Waterfield and J. Downward, Activation of phosphoinositide 3-kinase by interaction with Ras and by point mutation, *EMBO J.* **15** (1996) 2442–2451.
76. E. Rojas, V. Nassar-Gentina, M. Luxoro, M.E. Pollard and M.A. Carrasco, Inositol 1,4,5-trisphosphate-induced Ca^{2+} release from the sarcoplasmic reticulum and contraction in crustacean muscle, *Can. J. Physiol. Pharmacol.* **65** (1987) 672–680.
77. L. Huang, F. Hofer, G.S. Martin and S.H. Kim, Structural basis for the interaction of Ras with RalGDS, *Nat. Struct. Biol.* **5** (1998) 422–426.

78. C.P. Ponting, Novel domains in NADPH oxidase subunits, sorting nexins and PtdIns 3-kinases: binding partners of SH3 domains? *Protein Sci.* **5** (1996) 2353–2357.
79. E. Walker, O. Perisic, C. Ried, L. Stephens and R. Williams, Structural insights into phosphoinositide 3-kinase catalysis and signalling, *Nature* **402** (1999) 313–320.
80. J. Rizo and T.C. Sudhof, C2-domains, structure and function of a universal Ca^{2+}-binding domain, *J. Biol. Chem.* **273** (1998) 15879–15882.
81. M. Katan and V.L. Allen, Modular PH and C2 domains in membrane attachment and other functions, *FEBS Lett.* **452** (1999) 36–40.
82. L.O. Essen, O. Perisic, D.E. Lynch, M. Katan and R.L. Williams, A ternary metal binding site in the C2 domain of phosphoinositide-specific phospholipase C-delta1, *Biochemistry* **36** (1997) 2753–2762.
83. L.R. Stephens, K.T. Hughes and R.F. Irvine, Pathway of phosphatidylinositol(3,4,5)-trisphosphate synthesis in activated neutrophils, *Nature* **351** (1991) 33–39.
84. B. Vanhaesebroeck, K. Higashi, C. Raven, M. Welham, S. Anderson, P. Brennan, S.G. Ward and M.D. Waterfield, Autophosphorylation of p110delta phosphoinositide 3-kinase: a new paradigm for the regulation of lipid kinases in vitro and in vivo, *EMBO J.* **18** (1999) 1292–1302.
85. P. Hu, B. Margolis, E.Y. Skolnik, R. Lammers, A. Ullrich and J. Schlessinger, Interaction of phosphatidylinositol 3-kinase-associated p85 with epidermal growth factor and platelet-derived growth factor receptors, *Mol. Cell. Biol.* **12** (1992) 981–990.
86. C.J. McGlade, C. Ellis, M. Reedijk, D. Anderson, G. Mbamalu, A.D. Reith, G. Panayotou, P. End, A. Bernstein and A. Kazlauskas, SH2 domains of the p85 alpha subunit of phosphatidylinositol 3-kinase regulate binding to growth factor receptors, *Mol. Cell. Biol.* **12** (1992) 991–997.
87. A. Klippel, J.A. Escobedo, W.J. Fantl and L.T. Williams, The C-terminal SH2 domain of p85 accounts for the high affinity and specificity of the binding of phosphatidylinositol 3-kinase to phosphorylated platelet-derived growth factor beta receptor, *Mol. Cell. Biol.* **12** (1992) 1451–1459.
88. G. Panayotou, B. Bax, I. Gout, M. Federwisch, B. Wroblowski, R. Dhand, M.J. Fry, T.L. Blundell, A. Wollmer and M.D. Waterfield, Interaction of the p85 subunit of PI 3-kinase and its N-terminal SH2 domain with a PDGF receptor phosphorylation site: structural features and analysis of conformational changes, *EMBO J.* **11** (1992) 4261–4272.
89. S.E. Shoelson, M. Sivaraja, K.P. Williams, P. Hu, J. Schlessinger and M.A. Weiss, Specific phosphopeptide binding regulates a conformational change

in the PI 3-kinase SH2 domain associated with enzyme activation, *EMBO J.* **12** (1993) 795–802.

90. M.J. Layton, A.G. Harpur, G. Panayotou, P.I. Bastiaens and M.D. Waterfield, Binding of a diphosphotyrosine-containing peptide that mimics activated platelet-derived growth factor receptor beta induces oligomerization of phosphatidylinositol 3-kinase, *J. Biol. Chem.* **273** (1998) 33379–33385.
91. T. Rordorf-Nikolic, D.J. Van Horn, D. Chen, M.F. White and J.M. Backer, *J. Biol. Chem.* **270** (1995) 3662–3666.
92. D.R. Mattoon, B. Lamothe, I. Lax and J. Schlessinger, The docking protein Gab1 is the primary mediator of EGF-stimulated activation of the PI-3K/Akt cell survival pathway, *BMC Biol.* **2** (2004) 24.
93. M. Olayioye, R. Neve, H. Lane and H. Ne, The ErbB signaling network: receptor heterodimerization in development and cancer, *EMBO J.* **19** (2000) 3159–3167.
94. J.M. Backer, M.G. Myers, Jr., S.E. Shoelson, D.J. Chin, X.J. Sun, M. Miralpeix, P. Hu, B. Margolis, E.Y. Skolnik, J. Schlessinger, *et al.*, Phosphatidylinositol 3′-kinase is activated by association with IRS-1 during insulin stimulation, *EMBO J.* **11** (1992) 3469–3479.
95. M.L. Standaert, M.P. Sajan, A. Miura, G. Bandyopadhyay and R.V. Farese, Requirements for pYXXM motifs in Cbl for binding to the p85 subunit of phosphatidylinositol 3-kinase and Crk, and activation of atypical protein kinase C and glucose transport during insulin action in 3T3/L1 adipocytes, *Biochemistry* **43** (2004) 15494–15502.
96. J. Yu, C. Wjasow and J.M. Backer, Regulation of the p85/p110alpha phosphatidylinositol 3′-kinase. Distinct roles for the n-terminal and c-terminal SH2 domains, *J. Biol. Chem.* **273** (1998) 30199–30203.
97. C.E. Rudd and H. Schneider, Unifying concepts in CD28, ICOS and CTLA4 co-receptor signalling, *Nat. Rev. Immunol.* **3** (2003) 544–556.
98. Y. Xu, K.W. Harder, N.D. Huntington, M.L. Hibbs and D.M. Tarlinton, Lyn tyrosine kinase: accentuating the positive and the negative, *Immunity* **22** (2005) 9–18.
99. P.L. Schwartzberg, L.D. Finkelstein and J.A. Readinger, TEC-family kinases: regulators of T-helper-cell differentiation, *Nat. Rev. Immunol.* **5** (2005) 284–295.
100. L.A. Pitcher and N.S. van Oers, T-cell receptor signal transmission: who gives an ITAM? *Trends Immunol.* **24** (2003) 554–560.
101. N. Isakov, Immunoreceptor tyrosine-based activation motif (ITAM), a unique module linking antigen and Fc receptors to their signaling cascades, *J. Leukoc. Biol.* **61** (1997) 6–16.

102. N. Isakov, ITIMs and ITAMs. The Yin and Yang of antigen and Fc receptor-linked signaling machinery, *Immunol. Res.* **16** (1997) 85–100.
103. M. Reyes-Reyes, N. Mora, A. Zentella and C. Rosales, Phosphatidylinositol 3-kinase mediates integrin-dependent NF-kappaB and MAPK activation through separate signaling pathways, *J. Cell Sci.* **114** (2001) 1579–1589.
104. M.J. Humphries, Integrin structure, *Biochem. Soc. Trans.* **28** (2000) 311–339.
105. A. Khwaja, P. Rodriguez-Viciana, S. Wennstrom, P.H. Warne and J. Downward, Matrix adhesion and ras transformation both activate a phosphoinositide 3-OH kinase and protein kinase B/Akt cellular survival pathway, *EMBO J.* **16** (1997) 2783–2793.
106. H. Kurosu, T. Maehama, T. Okada, T. Yamamoto, S. Hoshino, Y. Fukui, M. Ui, O. Hazeki and T. Katada, Heterodimeric phosphoinositide 3-kinase consisting of p85 and p110beta is synergistically activated by the betagamma subunits of G proteins and phosphotyrosyl peptide, *J. Biol. Chem.* **272** (1997) 24252–24256.
107. S. Roche, J. Downward, P. Raynal and S.A. Courtneidge, A function for phosphatidylinositol 3-kinase beta (p85alpha-p110beta) in fibroblasts during mitogenesis: requirement for insulin- and lysophosphatidic acid-mediated signal transduction, *Mol. Cell. Biol.* **18** (1998) 7119–7129.
108. H. Daub, F.U. Weiss, C. Wallasch and A. Ullrich, Role of transactivation of the EGF receptor in signalling by G-protein-coupled receptors, *Nature* **379** (1996) 557–560.
109. M. Laffargue, P. Raynal, A. Yart, C. Peres, R. Wetzker, S. Roche, B. Payrastre and H. Chap, An epidermal growth factor receptor/Gab1 signaling pathway is required for activation of phosphoinositide 3-kinase by lysophosphatidic acid, *J. Biol. Chem.* **274** (1999) 32835–32841.
110. C. Czupalla, M. Culo, E.C. Muller, C. Brock, H.P. Reusch, K. Spicher, E. Krause and B. Nurnberg, Identification and characterization of the autophosphorylation sites of phosphoinositide 3-kinase isoforms β and γ, *J. Biol. Chem.* **278** (2003) 11536–11545.
111. M.E. Pacold, S. Suire, O. Perisic, S. Lara-Gonzalez, C.T. Davis, E.H. Walker, P.T. Hawkins, L. Stephens, J.F. Eccleston, R.L. Williams, Crystal structure and functional analysis of Ras binding to its effector phosphoinositide 3-kinase gamma, *Cell* **103** (2000) 931–943.
112. S. Krugmann, P.T. Hawkins, N. Pryer and S. Braselmann, Characterizing the interactions between the two subunits of the p101/p110gamma phosphoinositide 3-kinase and their role in the activation of this enzyme by G beta gamma subunits, *J. Biol. Chem.* **274** (1999) 17152–17158.
113. M.A. Andrade, C. Petosa, S.I. O'Donoghue, C.W. Muller and P. Bork, Comparison of ARM and HEAT protein repeats, *J. Mol. Biol.* **309** (2001) 1–18.

114. C. Czupalla, M. Culo, E.C. Muller, C. Brock, H.P. Reusch, K. Spicher, E. Krause and B. Nurnberg, Identification and characterization of the autophosphorylation sites of phosphoinositide 3-kinase isoforms beta and gamma, *J. Biol. Chem.* **278** (2003) 11536–11545.
115. C. Brock, M. Schaefer, H.P. Reusch, C. Czupalla, M. Michalke, K. Spicher, G. Schultz and B. Nurnberg, Roles of G beta gamma in membrane recruitment and activation of p110 gamma/p101 phosphoinositide 3-kinase gamma, *J. Cell Biol.* **160** (2003) 89–99.
116. B. Stoyanov, S. Volinia, T. Hanck, I. Rubio, M. Loubtchenkov, D. Malek, S. Stoyanova, B. Vanhaesebroeck, R. Dhand, B. Nurnberg, P. Gierschik, K. Seedorf, J.J. Hsuan, M.D. Waterfield and R. Wetzker, Cloning and characterisation of a G protein-activated human phosphatidlinositol 3-kinase, *Science* **269** (1995) 690–693.
117. K.R. Kerchner, R.L. Clay, G. McCleery, N. Watson, W.E. McIntire, C.S. Myung and J.C. Garrison, Differential sensitivity of phosphatidylinositol 3-kinase p110gamma to isoforms of G protein betagamma dimers, *J. Biol. Chem.* **279** (2004) 44554–44562.
118. S. Suire, J. Coadwell, G.J. Ferguson, K. Davidson, P. Hawkins and L. Stephens, p84, a new Gbetagamma-activated regulatory subunit of the type IB phosphoinositide 3-kinase p110gamma, *Curr. Biol.* **15** (2005) 566–570.
119. C. Murga, L. Laguinge, R. Wetzker, A. Cuadrado and J.S. Gutkind, Activation of Akt/protein kinase B by G protein-coupled receptors. A role for alpha and beta gamma subunits of heterotrimeric G proteins acting through phosphatidylinositol-3-OH kinasegamma, *J. Biol. Chem.* **273** (1998) 19080–19085.
120. T. Bondeva, L. Pirola, G. Bulgarelli-Leva, I. Rubio, R. Wetzker and M. Wymann, Bifurcation of lipid and protein kinase signals of PI3Kγ to the protein kinases PBK and MAPK, *Science* **282** (1998) 293–296.
121. U. Maier, A. Babich and B. Nurnberg, Roles of non-catalytic subunits in gbetagamma-induced activation of class I phosphoinositide 3-kinase isoforms beta and gamma, *J. Biol. Chem.* **274** (1999) 29311–29317.
122. A. Bondev, I. Rubio and R. Wetzker, Differential regulation of lipid and protein kinase activities of phosphoinositide 3-kinase gamma in vitro, *Biol. Chem.* **380** (1999) 1337–1340.
123. J. Domin, I. Gaidarov, M. Smith, J. Keen and M. Waterfield, The class II phosphoinositide 3-kinase PI3K-C2alpha is concentrated in the trans-Golgi network and present in clathrin-coated vesicles, *J. Biol. Chem.* **275** (2000) 11943–11950.

124. C.D. Ellson, S. Gobert-Gosse, K.E. Anderson, K. Davidson, H. Erdjument-Bromage, P. Tempst, J.W. Thuring, M.A. Cooper, Z.Y. Lim, A.B. Holmes, P.R. Gaffney, J. Coadwell, E.R. Chilvers, P.T. Hawkins and L.R. Stephens, PtdIns(3)P regulates the neutrophil oxidase complex by binding to the PX domain of p40(phox), *Nat. Cell Biol.* **3** (2001) 679–682.
125. T.K. Sato, M. Overduin and S.D. Emr, Location, location, location: membrane targeting directed by PX domains, *Science* **294** (2001) 1881–1885.
126. C.D. Ellson, S. Andrews, L.R. Stephens and P.T. Hawkins, The PX domain: a new phosphoinositide-binding module, *J. Cell Sci.* **115** (2002) 1099–1105.
127. T. Ago, R. Takeya, H. Hiroaki, F. Kuribayashi, T. Ito, D. Kohda and H. Sumimoto, The PX domain as a novel phosphoinositide-binding module, *Biochem. Biophys. Res. Commun.* **287** (2001) 733–738.
128. X. Song, W. Xu, A. Zhang, G. Huang, X. Liang, J.V. Virbasius, M.P. Czech and G.W. Zhou, Phox homology domains specifically bind phosphatidylinositol phosphates, *Biochemistry* **40** (2001) 8940–8944.
129. Y. Zhan, J.V. Virbasius, X. Song, D.P. Pomerleau and G.W. Zhou, The p40phox and p47phox PX domains of NADPH oxidase target cell membranes via direct and indirect recruitment by phosphoinositides, *J. Biol. Chem.* **277** (2002) 4512–4518.
130. M. Wheeler and J. Domin, Recruitment of the class II phosphoinositide 3-kinase C2beta to the epidermal growth factor receptor: role of Grb2, *Mol. Cell. Biol.* **21** (2001) 6660–6667.
131. A. Arcaro, M.J. Zvelebil, C. Wallasch, A. Ullrich, M.D. Waterfield and J. Domin, Class II phosphoinositide 3-kinase are downstream targets of activated polypeptide growth factor receptors, *Mol. Cell. Biol.* **20** (2000) 3817–3830.
132. H. Banfic, X. Tang, I.H. Batty, C.P. Downes, C. Chen and S.E. Rittenhouse, A novel integrin-activated pathway forms PKB/Akt-stimulatory phosphatidylinositol 3,4-bisphosphate via phosphatidylinositol 3-phosphate in platelets, *J. Biol. Chem.* **273** (1998) 13–16.
133. J. Zhang, H. Banfic, F. Straforini, L. Tosi, S. Volinia and S. Rittenhouse, A type II phosphoinositide 3-kinase is stimulated via activated integrin in platelets. A source of phosphatidylinositol 3-phosphate, *J. Biol. Chem.* **273** (1998) 14081–14084.
134. A. Sindic, A. Aleksandrova, A.P. Fields, S. Volinia and H. Banfic, Presence and activation of nuclear phosphoinositide 3-kinase C2beta during compensatory liver growth, *J. Biol. Chem.* **276** (2001) 17754–17761.
135. D. Visnjic, J. Curic, V. Crljen, D. Batinic, S. Volinia and H. Banfic, Nuclear phosphoinositide 3-kinase C2beta activation during G2/M phase of the cell cycle in HL-60 cells, *Biochim. Biophys. Acta* **1631** (2003) 61–71.

136. V. Crljen, S. Volinia and H. Banfic, Hepatocyte growth factor activates phosphoinositide 3-kinase C2beta in renal brush-border plasma membranes, *Biochem. J.* **365** (2002) 791–799.
137. A. Arcaro, U.K. Khanzada, B. Vanhaesebroeck, T.D. Tetley, M.D. Waterfield and M.J. Seckl, Two distinct phosphoinositide 3-kinases mediate polypeptide growth factor-stimulated PKB activation, *EMBO J.* **21** (2002) 5097–5108.
138. I. Gaidarov, M.E. Smith, J. Domin and J.H. Keen, The class II phosphoinositide 3-kinase C2alpha is activated by clathrin and regulates clathrin-mediated membrane trafficking, *Mol. Cell* **7** (2001) 443–449.
139. R.A. Brown, J. Domin, A. Arcaro, M.D. Waterfield and P.R. Shepherd, Insulin activates the alpha isoform of class II phosphoinositide 3-kinase, *J. Biol. Chem.* **274** (1999) 14529–14532.
140. S.J. Turner, J. Domin, M.D. Waterfield, S.G. Ward and J. Westwick, The CC chemokine monocyte chemotactic peptide-1 activates both the class I p85/p110 Phosphatidylinositol 3-kinase and the class II PI3K-C2α, *J. Biol. Chem.* **273** (1998) 25987–25995.
141. L.K. MacDougall, M.E. Gagou, S.J. Leevers, E. Hafen and M.D. Waterfield, Targeted expression of the class II phosphoinositide 3-kinase in *Drosophila melanogaster* reveals lipid kinase-dependent effects on patterning and interactions with receptor signaling pathways, *Mol. Cell. Biol.* **24** (2004) 796–808.
142. S.S. El Sheikh, J. Domin, P. Tomtitchong, P. Abel, G. Stamp and E.N. Lalani, Topographical expression of class IA and class II phosphoinositide 3-kinase enzymes in normal human tissues is consistent with a role in differentiation, *BMC Clin. Pathol.* **3** (2003) 4.
143. S. Kang, J. Song, J. Kang, H. Kang, D. Lee, Y. Lee and D. Park, Suppression of the alpha-isoform of class II phosphoinositide 3-kinase gene expression leads to apoptotic cell death, *Biochem. Biophys. Res. Commun.* **329** (2005) 6–10.
144. S.S. El Sheikh, J. Domin, P. Abel, G. Stamp and E. Lalani, Androgen-independent prostate cancer: potential role of androgen and ErbB receptor signal transduction crosstalk, *Neoplasia* **5** (2003) 99–109.
145. V.A. Bankaitis, L.M. Johnson and S.D. Emr, Isolation of yeast mutants defective in protein targeting to the vacuole, *Proc. Natl. Acad. Sci. USA* **83** (1986) 9075–9079.
146. P.K. Herman and S.D. Emr, Characterization of VPS34, a gene required for vacuolar protein sorting and vacuole segregation in Saccharomyces cerevisiae, *Mol. Cell. Biol.* **10** (1990) 6742–6754.
147. Y.V. Budovskaya, H. Hama, D.B. DeWald and P.K. Herman, The C terminus of the Vps34p phosphoinositide 3-kinase is necessary and sufficient for

the interaction with the Vps15p protein kinase, *J. Biol. Chem.* **277** (2002) 287–294.

148. P.K. Herman, J.H. Stack, J.A. DeModena and S.D. Emr, A novel protein kinase homolog essential for protein sorting to the yeast lysosome-like vacuole, *Cell* **64** (1991) 425–437.
149. C. Panaretou, J. Domin, S. Cockcroft and M.D. Waterfield, Characterization of p150, an adaptor protein for the human phosphatidylinositol (PtdIns) 3-kinase. Substrate presentation by phosphatidylinositol transfer protein to the p150.Ptdins 3-kinase complex, *J. Biol. Chem.* **272** (1997) 2477–2485.
150. J.T. Murray, C. Panaretou, H. Stenmark, M. Miaczynska and J.M. Backer, Role of Rab5 in the recruitment of hVps34/p150 to the early endosome, *Traffic* **3** (2002) 416–427.
151. P.K. Herman, J.H. Stack and S.D. Emr, A genetic and structural analysis of yeast vps15 protein kinase: evidence for a direct role of vps15p in vacuolar protein delivery, *EMBO J.* **10** (1991) 4049–4060.
152. P. De Camilli, S.D. Emr, P.S. McPherson and P. Novick, Phosphoinositides as regulators in membrane traffic, *Science* **271** (1996) 1533–1539.
153. T.K. Sato, T. Darsow and S.D. Emr, Vam7p, a SNAP-25-like molecule, and Vam3p, a syntaxin homolog, function together in yeast vacuolar protein trafficking, *Mol. Cell. Biol.* **18** (1998) 5308–5319.
154. C.G. Burd, M. Peterson, C.R. Cowles and S.D. Emr, A novel Sec18p/NSF-dependent complex required for Golgi-to-endosome transport in yeast, *Mol. Biol. Cell* **8** (1997) 1089–1104.
155. C.G. Burd and S.D. Emr, Phosphatidylinositol (3)-phosphate signalling mediated by specific binding to RING FYVE domains, *Mol. Cell* **2** (1998) 157–162.
156. M.L. Cheever, T.K. Sato, B.T. de, T.G. Kutateladze, S.D. Emr and M. Overduin, Phox domain interaction with PtdIns(3)P targets the Vam7 t-SNARE to vacuole membranes, *Nat. Cell Biol.* **3** (2001) 613–618.
157. A. Kihara, T. Noda, N. Ishihara and Y. Ohsumi, Two distinct Vps34 phosphatidylinositol 3-kinase complexes function in autophagy and carboxypeptidase Y sorting in *Saccharomyces cerevisiae*, *J. Cell Biol.* **152** (2001) 519–530.
158. J.A. Kiel, K.B. Rechinger, D.K. van, I, F.A. Salomons, V.I. Titorenko and M. Veenhuis, The Hansenula polymorpha PDD1 gene product, essential for the selective degradation of peroxisomes, is a homologue of *Saccharomyces cerevisiae* Vps34p, *Yeast* **15** (1999) 741–754.
159. W.A. Dunn, Jr., Studies on the mechanisms of autophagy: maturation of the autophagic vacuole, *J. Cell Biol.* **110** (1990) 1935–1945.

160. A. Petiot, E. Ogier-Denis, E.F. Blommaart, A.J. Meijer and P. Codogno, Distinct classes of phosphatidylinositol 3′-kinases are involved in signaling pathways that control macroautophagy in HT-29 cells, *J. Biol. Chem.* **275** (2000) 992–998.
161. A.E. Wurmser and S.D. Emr, Novel PtdIns(3)P-binding protein Etf1 functions as an effector of the Vps34 PtdIns 3-kinase in autophagy, *J. Cell Biol.* **158** (2002) 761–772.
162. P. Mitra, Y. Zhang, L.E. Rameh, M.P. Ivshina, D. McCollum, J.J. Nunnari, G.M. Hendricks, M.L. Kerr, S.J. Field, L.C. Cantley and A.H. Ross, A novel phosphatidylinositol(3,4,5)P3 pathway in fission yeast, *J. Cell Biol.* **166** (2004) 205–211.
163. Y. Zhang, R. Sugiura, Y. Lu, M. Asami, T. Maeda, T. Itoh, T. Takenawa, H. Shuntoh and T. Kuno, Phosphatidylinositol 4-phosphate 5-kinase Its3 and calcineurin Ppb1 coordinately regulate cytokinesis in fission yeast, *J. Biol. Chem.* **275** (2000) 35600–35606.
164. D.B. DeWald, Phosphoinositide signaling: vac to the future in fab1 kinase regulation, *Curr. Biol.* **12** (2002) R491–R492.
165. D.P. Ten and C.H. Heldin, Signal transduction. An anchor for activation, *Nature* **397** (1999) 109–111.
166. M. Aoki, C. Schetter, M. Himly, O. Batista, H.W. Chang and P.K. Vogt, The catalytic subunit of phosphoinositide 3-kinase: requirements for oncogenicity, *J. Biol. Chem.* **275** (2000) 6267–6275.
167. A. Klippel, M.A. Escobedo, M.S. Wachowicz, G. Apell, T.W. Brown, M.A. Giedlin, W.M. Kavanaugh and L.T. Williams, Activation of phosphatidylinositol 3-kinase is sufficient for cell cycle entry and promotes cellular changes characteristic of oncogenic transformation, *Mol. Cell. Biol.* **18** (1998) 5699–5711.
168. J.H. Stack, P.K. Herman, P.V. Schu and S.D. Emr, A membrane associated complex containing the vps15 protein kinase and the vps34 PI 3-kinase is essential for protein sorting to the yeast lysosome-like vacuole, *EMBO J.* **12** (1993) 2195–2204.
169. M. Wheeler and J. Domin, The N-terminus of phosphoinositide 3-kinase-C2 beta regulates lipid kinase activity and binding to clathrin, *J. Cell. Physiol.* **206** (2006) 586–593.
170. A.M. Martelli, G. Tabellini, P. Borgatti, R. Bortul, S. Capitani and L.M. Neri, Nuclear lipids: new functions for old molecules? *J. Cell. Biochem.* **88** (2003) 455–461.
171. R. Chen, V.H. Kang, J. Chen, J.C. Shope, J. Torabinejad, D.B. DeWald and G.D. Prestwich, A monoclonal antibody to visualize PtdIns(3,4,5)P(3) in cells, *J. Histochem. Cytochem.* **50** (2002) 697–708.

172. S.L. Osborne, C.L. Thomas, S. Gschmeissner and G. Schiavo, Nuclear PtdIns(4,5)P2 assembles in a mitotically regulated particle involved in pre-mRNA splicing, *J. Cell Sci.* **114** (2001) 2501–2511.
173. S.A. Watt, G. Kular, I.N. Fleming, C.P. Downes and J.M. Lucocq, Subcellular localization of phosphatidylinositol 4,5-bisphosphate using the pleckstrin homology domain of phospholipase C delta1, *Biochem. J.* **363** (2002) 657–666.
174. M.A. Lemmon and K.M. Ferguson, Signal-dependent membrane targeting by pleckstrin homology (PH) domains, *Biochem. J.* **350**(Pt 1) (2000) 1–18.
175. L.M. Neri, A.M. Martelli, P. Borgatti, M.L. Colamussi, M. Marchisio and S. Capitani, Increase in nuclear phosphatidylinositol 3-kinase activity and phosphatidylinositol (3,4,5) trisphosphate synthesis precede PKC-zeta translocation to the nucleus of NGF-treated PC12 cells, *FASEB J.* **13** (1999) 2299–2310.
176. W.O. White, M.L. Seibenhener and M.W. Wooten, Phosphorylation of tyrosine 256 facilitates nuclear import of atypical protein kinase C, *J. Cell. Biochem.* **85** (2002) 42–53.
177. K. Ye, K.J. Hurt, F.Y. Wu, M. Fang, H.R. Luo, J.J. Hong, S. Blackshaw, C.D. Ferris and S.H. Snyder, Pike. A nuclear gtpase that enhances PI3kinase activity and is regulated by protein 4.1N, *Cell* **103** (2000) 919–930.
178. A. Metjian, R.L. Roll, A.D. Ma and C.S. Abrams, Agonists cause nuclear translocation of phosphatidylinositol 3-kinase gamma. A Gbetagamma-dependent pathway that requires the p110gamma amino terminus, *J. Biol. Chem.* **274** (1999) 27943–27947.
179. S.A. Didichenko and M. Thelen, Phosphatidylinositol 3-kinase c2alpha contains a nuclear localization sequence and associates with nuclear speckles, *J. Biol. Chem.* **276** (2001) 48135–48142.
180. T.D. Bunney, P.A. Watkins, A.F. Beven, P.J. Shaw, L.E. Hernandez, G.P. Lomonossoff, M. Shanks, J. Peart and B.K. Drobak, Association of phosphatidylinositol 3-kinase with nuclear transcription sites in higher plants, *Plant Cell* **12** (2000) 1679–1688.
181. O.J. Rando, K. Zhao, P. Janmey and G.R. Crabtree, Phosphatidylinositol-dependent actin filament binding by the SWI/SNF-like BAF chromatin remodeling complex, *Proc. Natl. Acad. Sci. USA* **99** (2002) 2824–2829.
182. T. Yokogawa, S. Nagata, Y. Nishio, T. Tsutsumi, S. Ihara, R. Shirai, K. Morita, M. Umeda, Y. Shirai, N. Saitoh and Y. Fukui, Evidence that 3′-phosphorylated polyphosphoinositides are generated at the nuclear surface: use of immunostaining technique with monoclonal antibodies specific for PI 3,4-P(2), *FEBS Lett.* **473** (2000) 222–226.

183. S.B. Helliwell, P. Wagner, J. Kunz, M. uter-Reinhard, R. Henriquez and M.N. Hall, TOR1 and TOR2 are structurally and functionally similar but not identical phosphatidylinositol kinase homologues in yeast, *Mol. Biol. Cell* **5** (1994) 105–118.
184. D.M. Sabatini, H. Erdjument Bromage, M. Lui, P. Tempst and S.H. Snyder, RAFT1: a mammalian protein that binds to FKBP12 in a rapamycin-dependent fashion and is homologous to yeast TORs, *Cell* **78** (1994) 35–43.
185. T.E. Harris and J.C. Lawrence, Jr., TOR signaling, *Sci. STKE* **2003** (2003) 15.
186. S.J. Collis, T.L. DeWeese, P.A. Jeggo and A.R. Parker, The life and death of DNA-PK, *Oncogene* **24** (2005) 949–961.
187. P.J. McKinnon, ATM and ataxia telangiectasia, *EMBO Rep.* **5** (2004) 772–776.
188. J.M. Gaullier, A. Simonsen, A. D'Arrigo, B. Bremnes, H. Stenmark and R. Aasland, FYVE fingers bind PtdIns(3)P, *Nature* **394** (1998) 432–433.
189. H. Stenmark, R. Aasland and P.C. Driscoll, The phosphatidylinositol 3-phosphate-binding FYVE finger, *FEBS Lett.* **513** (2002) 77–84.
190. P.C. Driscoll and A.L. Vuidepot, Peripheral membrane proteins: FYVE sticky fingers, *Curr. Biol.* **9** (1999) R857–R860.
191. D.J. Gillooly, I.C. Morrow, M. Lindsay, R. Gould, N.J. Bryant, J.M. Gaullier, R.G. Parton and H. Stenmark, Localization of phosphatidylinositol 3-phosphate in yeast and mammalian cells, *EMBO J.* **19** (2000) 4577–4588.
192. B.J. Mayer, R. Ren, K.L. Clark and D. Baltimore, A putative modular domain present in diverse signaling proteins, *Cell* **73** (1993) 629–630.
193. R.J. Haslam, H.B. Koide and B.A. Hemmings, Pleckstrin domain homology, *Nature* **363** (1993) 309–310.
194. A. Musacchio, T. Gibson, P. Rice, J. Thompson and M. Saraste, The PH domain: a common piece in the structural patchwork of signalling proteins, *Trends Biochem. Sci.* **18** (1993) 343–348.
195. J.E. Harlan, P.J. Hajduk, H.S. Yoon and S.W. Fesik, Pleckstrin homology domains bind to phosphatidylinositol-4,5-bisphosphate, *Nature* **371** (1994) 168–170.
196. A. Klippel, W.M. Kavanaugh, D. Pot and L.T. Williams, A specific product of phosphatidylinositol3-kinase directly activates the protein kinase Akt through its plekstrin homology domain, *Mol. Cell. Biol.* **17** (1997) 338–344.
197. B.A. Hemmings, Akt signaling: linking membrane events to life and death decisions, *Science* **275** (1997) 628–630.

198. C.C. Milburn, M. Deak, S.M. Kelly, N.C. Price, D.R. Alessi and D.M. van Aalten, Binding of phosphatidylinositol 3,4,5-trisphosphate to the pleckstrin homology domain of protein kinase B induces a conformational change, *Biochem. J.* **375** (2003) 531–538.
199. A. Safi, M. Vandromme, S. Caussanel, L. Valdacci, D. Baas, M. Vidal, G. Brun, L. Schaeffer and E. Goillot, Role for the pleckstrin homology domain-containing protein CKIP-1 in phosphatidylinositol 3-kinase-regulated muscle differentiation, *Mol. Cell. Biol.* **24** (2004) 1245–1255.
200. D.B. van Rossum, R.L. Patterson, S. Sharma, R.K. Barrow, M. Kornberg, D.L. Gill and S.H. Snyder, Phospholipase Cgamma1 controls surface expression of TRPC3 through an intermolecular PH domain, *Nature* **434** (2005) 99–104.
201. A.W. Segal, The NADPH oxidase and chronic granulomatous disease, *Mol. Med. Today* **2** (1996) 129–135.
202. B.K. Kay, M. Yamabhai, B. Wendland and S.D. Emr, Identification of a novel domain shared by putative components of the endocytic and cytoskeletal machinery, *Protein Sci.* **8** (1999) 435–438.
203. V. Legendre-Guillemin, S. Wasiak, N.K. Hussain, A. Angers and P.S. McPherson, ENTH/ANTH proteins and clathrin-mediated membrane budding, *J. Cell Sci.* **117** (2004) 9–18.
204. J.A. Rosenthal, H. Chen, V.I. Slepnev, L. Pellegrini, A.E. Salcini, P.P. Di Fiore and C.P. De, The epsins define a family of proteins that interact with components of the clathrin coat and contain a new protein module, *J. Biol. Chem.* **274** (1999) 33959–33965.
205. T. Itoh, S. Koshiba, T. Kigawa, A. Kikuchi, S. Yokoyama and T. Takenawa, Role of the ENTH domain in phosphatidylinositol-4,5-bisphosphate binding and endocytosis, *Science* **291** (2001) 1047–1051.
206. M.G. Ford, B.M. Pearse, M.K. Higgins, Y. Vallis, D.J. Owen, A. Gibson, C.R. Hopkins, P.R. Evans and H.T. McMahon, Simultaneous binding of PtdIns(4,5)P2 and clathrin by AP180 in the nucleation of clathrin lattices on membranes, *Science* **291** (2001) 1051–1055.
207. T.S. Hyun, D.S. Rao, D. Saint-Dic, L.E. Michael, P.D. Kumar, S.V. Bradley, I.F. Mizukami, K.I. Oravecz-Wilson and T.S. Ross, HIP1 and HIP1r stabilize receptor tyrosine kinases and bind 3-phosphoinositides via epsin N-terminal homology domains, *J. Biol. Chem.* **279** (2004) 14294–14306.
208. S. Friant, E.I. Pecheur, A. Eugster, F. Michel, Y. Lefkir, D. Nourrisson and F. Letourneur, Ent3p Is a PtdIns(3,5)P2 effector required for protein sorting to the multivesicular body, *Dev. Cell* **5** (2003) 499–511.

209. Y. Xu, L.F. Seet, B. Hanson and W. Hong, The Phox homology (PX) domain, a new player in phosphoinositide signalling, *Biochem. J.* **360** (2001) 513–530.
210. A. Shisheva, T.C. Sudhof and M.P. Czech, Cloning, characterization, and expression of a novel GDP dissociation inhibitor isoform from skeletal muscle, *Mol. Cell. Biol.* **14** (1994) 3459–3468.
211. G. Odorizzi, M. Babst and S.D. Emr, Phosphoinositide signaling and the regulation of membrane trafficking in yeast, *Trends Biochem. Sci.* **25** (2000) 229–235.
212. G. Odorizzi, M. Babst and S.D. Emr, Fab1p PtdIns(3)P 5-kinase function essential for protein sorting in the multivesicular body, *Cell* **95** (1998) 847–858.
213. M.R. Peterson, C.G. Burd and S.D. Emr, Vac1p coordinates Rab and phosphatidylinositol 3-kinase signaling in Vps45p-dependent vesicle docking/fusion at the endosome, *Curr. Biol.* **9** (1999) 159–162.
214. D.J. Katzmann, C.J. Stefan, M. Babst and S.D. Emr, Vps27 recruits ESCRT machinery to endosomes during MVB sorting, *J. Cell Biol.* **162** (2003) 413–423.
215. F. Itoh, N. Divecha, L. Brocks, L. Oomen, H. Janssen, J. Calafat, S. Itoh and P.P. Dijke, The FYVE domain in Smad anchor for receptor activation (SARA) is sufficient for localization of SARA in early endosomes and regulates TGF-beta/Smad signalling, *Genes Cells* **7** (2002) 321–331.
216. S.K. Dove, R.K. McEwen, F.T. Cooke, P.J. Parker and R.H. Michell, Phosphatidylinositol 3,5-bisphosphate: a novel lipid that links stress responses to membrane trafficking events 1, *Biochem. Soc. Trans.* **27** (1999) 674–677.
217. R.H. Michell, V.T. Heath, M.A. Lemmon and S.K. Dove, Phosphatidyinositol 3,5-bisphosphate: metabolism and cellular functions, *Trends Biochem. Sci.* **31** (2006) 52–63.
218. A. Eugster, E.I. Pecheur, F. Michel, B. Winsor, F. Lefourneur and S. Friant, Ent5p is required with Ent3p and vps27p for ubiquitin-dependent protein sorting into the multivesicular body, *Mol. Biol. Cell* **15** (2004) 3031–3041.
219. S.K. Dove, R.C. Piper, R.K. McEwen, J.W. Yu, M.C. King, D.C. Hughes, J. Thuring, A.B. Holmes, F.T. Cooke, R.H. Michell, P.J. Parker and M.A. Lemmon, Svp1p defines a family of phosphatidylinositol 3,5-bisphosphate effectors, *EMBO J.* **23** (2004) 1922–1933.
220. T.R. Jeffries, S.K. Dove, R.H. Michell and P.J. Parker, PtdIns-specific MPR pathway association of a novel WD40 repeat protein, WIPI49, *Mol. Biol. Cell* **15** (2004) 2652–2663.

221. A. Bellacosa, J.R. Testa, S.P. Staal and P.N. Tsichlis, A retroviral oncogene, akt, encoding a serine-threonine kinase containing an SH2-like region, *Science* **254** (1991) 274–277.
222. P.J. Coffer and J.R. Woodgett, Molecular cloning and characterisation of a novel putative protein-serine kinase related to the cAMP-dependent and protein kinase C families, *Eur. J. Biochem.* **201** (1991) 475–481.
223. P.F. Jones, T. Jakubowicz, F.J. Pitossi, F. Maurer and B.A. Hemmings, Molecular cloning and identification of a serine/threonine protein kinase of the second-messenger subfamily, *Proc. Natl. Acad. Sci. USA* **88** (1991) 4171–4175.
224. M. Cho, V. Villani and P. D'Eustachio, A linkage map of distal mouse chromosome 12, *Mamm. Genome* **1** (1991) 30–36.
225. S.P. Staal, K. Huebner, C.M. Croce, N.Z. Parse and J.R. Testa, The AKT1 proto-oncogene maps to human chromosome 14, bandq32, *Genomics* **2** (1988) 96–98.
226. J.Q. Cheng, A.K. Godwin, A. Bellacosa, T. Taguchi, T.F. Franke, T.C. Hamilton, P.N. Tsichlis and J.R. Testa, AKT2, a putative oncogene encoding a member of a subfamily of protein-serine/threonine kinases, is amplified in human ovarian carcinomas, *Proc. Natl. Acad. Sci. USA* **89** (1992) 9267–9271.
227. H. Konishi, S. Kuroda, M. Tanaka, H. Matsuzaki, Y. Ono, K. Kameyama, T. Haga and U. Kikkawa, Molecular cloning and characterization of a new member of the RAC protein kinase family: association of the pleckstrin homology domain of three types of RAC protein kinase with protein kinase C subspecies and beta gamma subunits of G proteins, *Biochem. Biophys. Res. Commun.* **216** (1995) 526–534.
228. J. Okano, I. Gaslightwala, M.J. Birnbaum, A.K. Rustgi and H. Nakagawa, Akt/protein kinase B isoforms are differentially regulated by epidermal growth factor stimulation, *J. Biol. Chem.* **275** (2000) 30934–30942.
229. S. Masure, B. Haefner, J.J. Wesselink, E. Hoefnagel, E. Mortier, P. Verhasselt, A. Tuytelaars, R. Gordon and A. Richardson, Molecular cloning, expression and characterization of the human serine/threonine kinase Akt-3, *Eur. J. Biochem.* **265** (1999) 353–360.
230. D.A. Altomare, G.E. Lyons, Y. Mitsuuchi, J.Q. Cheng and J.R. Testa, Akt2 mRNA is highly expressed in embryonic brown fat and the AKT2 kinase is activated by insulin, *Oncogene* **16** (1998) 2407–2411.
231. T.F. Franke, D.R. Kaplan, L.C. Cantley and A. Toker, Direct regulation of the Akt proto-oncogene product by phosphatidylinositol-3,4-bisphosphate, *Science* **275** (1997) 665–668.

232. T.F. Franke, S.I. Yang, T.O. Chan, K. Datta, A. Kazlauskas, D.K. Morrison, D.R. Kaplan and P.N. Tsichlis, The protein kinase encoded by the Akt proto-oncogene is a target of the PDGF-activated phosphatidylinositol 3-kinase, *Cell* **81** (1995) 727–736.
233. D.R. Alessi, M. Andjelkovic, B. Caudwell, P. Cron, N. Morrice, P. Cohen and B.A. Hemmings, Mechanism of activation of protein kinase B by insulin and IGF-1, *EMBO J.* **15** (1996) 6541–6551.
234. B.M. Burgering and P.J. Coffer, Protein kinase B (c-Akt) in phosphatidylinositol-3-OH kinase signal transduction, *Nature* **376** (1995) 599–602.
235. D.R. Alessi, S.R. James, C.P. Downes, A.B. Holmes, P.R.J. Gaffney, C.B. Reese and P. Cohen, Characterization of a 3-phosphoinositide-dependent protein kinase which phosphorylates and activates protein kinase Bα, *Curr. Biol.* **7** (1997) 261–269.
236. S.S. Castillo, J. Brognard, P.A. Petukhov, C. Zhang, J. Tsurutani, C.A. Granville, M. Li, M. Jung, K.A. West, J.G. Gills, A.P. Kozikowski and P.A. Dennis, Preferential inhibition of Akt and killing of Akt-dependent cancer cells by rationally designed phosphatidylinositol ether lipid analogues, *Cancer Res.* **64** (2004) 2782–2792.
237. E.A. Clark, W.G. King, J.S. Brugge, M. Symons and R.O. Hynes, Integrin-mediated signals regulated by members of the rho family of GTPases, *J. Cell Biol.* **142** (1998) 573–586.
238. M. Shaw, P. Cohen and D.R. Alessi, The activation of protein kinase B by H2O2 or heat shock is mediated by phosphoinositide 3-kinase and not by mitogen-activated protein kinase-activated protein kinase-2 1, *Biochem. J.* **336**(Pt 1) (1998) 241–246.
239. T. Nishino, C.D. Pusey and J. Domin, Elevated Akt phosphorylation as an indicator of renal tubular epithelial cell stress, *J. Biol. Chem.* **277** (2002) 33943–33949.
240. R. Meier, D.R. Alessi, P. Cron, M. Andjelkovic and B.A. Hemmings, Mitogenic activation, phosphorylation, and nuclear translocation of protein kinase Bbeta, *J. Biol. Chem.* **272** (1997) 30491–30497.
241. S.M. Maira, I. Galetic, D.P. Brazil, S. Kaech, E. Ingley, M. Thelen and B.A. Hemmings, Carboxyl-terminal modulator protein (CTMP), a negative regulator of PKB/Akt and v-Akt at the plasma membrane, *Science* **294** (2001) 374–380.
242. D. Stokoe, L.R. Stephens, T. Copeland, P.R. Gaffney, C.B. Reese, G.F. Painter, A.B. Holmes, F. McCormick and P.T. Hawkins, Dual role of phosphatidylinositol-3,4,5-trisphosphate in the activation of protein kinase B, *Science* **277** (1997) 567–570.

243. L. Stephens, K. Anderson, D. Stokoe, H. Erdjument-Bromage, G.F. Painter, A.B. Holmes, P.R. Gaffney, C.B. Reese, F. McCormick, P. Tempst, J. Coadwell and P.T. Hawkins, Protein kinase B kinases that mediate phosphatidylinositol 3,4,5-trisphosphate-dependent activation of protein kinase B, *Science* **279** (1998) 710–714.
244. A.D. Kohn, S.A. Summers, M.J. Birnbaum and R.A. Roth, Expression of a constitutively active Akt Ser/Thr kinase in 3T3-L1 adipocytes stimulates glucose uptake and glucose transporter 4 translocation, *J. Biol. Chem.* **271** (1996) 31372–31378.
245. M.A. Lim, C.K. Kikani, M.J. Wick and L.Q. Dong, Nuclear translocation of 3′-phosphoinositide-dependent protein kinase 1 (PDK-1): a potential regulatory mechanism for PDK-1 function, *Proc. Natl. Acad. Sci. USA* **100** (2003) 14006–14011.
246. H.J. Hinton, D.R. Alessi and D.A. Cantrell, The serine kinase phosphoinositide-dependent kinase 1 (PDK1) regulates T cell development, *Nat. Immunol.* **5** (2004) 539–545.
247. D. Komander, A. Fairservice, M. Deak, G.S. Kular, A.R. Prescott, D.C. Peter, S.T. Safrany, D.R. Alessi and D.M. van Aalten, Structural insights into the regulation of PDK1 by phosphoinositides and inositol phosphates, *EMBO J.* **23** (2004) 3918–3928.
248. M. Frodin, T.L. Antal, B.A. Dummler, C.J. Jensen, M. Deak, S. Gammeltoft and R.M. Biondi, A phosphoserine/threonine-binding pocket in AGC kinases and PDK1 mediates activation by hydrophobic motif phosphorylation, *EMBO J.* **21** (2002) 5396–5407.
249. X. Cheng, Y. Ma, M. Moore, B.A. Hemmings and S.S. Taylor, Phosphorylation and activation of cAMP-dependent protein kinase by phosphoinositide-dependent protein kinase, *Proc. Natl. Acad. Sci. USA* **95** (1998) 9849–9854.
250. M.M. Chou, W. Hou, J. Johnson, L.K. Graham, M.H. Lee, C.S. Chen, A.C. Newton, B.S. Schaffhausen, A. Toker, Regulation of protein kinase C zeta by PI 3-kinase and PDK-1, *Curr. Biol.* **8** (1998) 1069–1077.
251. N. Pullen, P.B. Dennis, M. Andjelkovic, A. Dufner, S.C. Kozma, B.A. Hemmings and G. Thomas, Phosphorylation and activation of p70s6k by PDK1, *Science* **279** (1998) 707–710.
252. C.J. Jensen, M.B. Buch, T.O. Krag, B.A. Hemmings, S. Gammeltoft and M. Frodin, 90-kDa ribosomal S6 kinase is phosphorylated and activated by 3-phosphoinositide-dependent protein kinase-1, *J. Biol. Chem.* **274** (1999) 27168–27176.
253. J. Park, M.L. Leong, P. Buse, A.C. Maiyar, G.L. Firestone and B.A. Hemmings, Serum and glucocorticoid-inducible kinase (SGK) is a target

of the PI 3-kinase-stimulated signaling pathway, *EMBO J.* **18** (1999) 3024–3033.

254. H. Harada, B. Becknell, M. Wilm, M. Mann, L.J. Huang, S.S. Taylor, J.D. Scott and S.J. Korsmeyer, Phosphorylation and inactivation of BAD by mitochondria-anchored protein kinase A, *Mol. Cell* **3** (1999) 413–422.
255. N. Terada, J.J. Lucas, A. Szepesi, R.A. Franklin, K. Takase and E.W. Gelfand, Rapamycin inhibits the phosphorylation of p70 S6 kinase in IL-2 and mitogen-activated human T cells, *Biochem. Biophys. Res. Commun.* **186** (1992) 1315–1321.
256. M. Wilson, A.R. Burt, G. Milligan and N.G. Anderson, Wortmannin-sensitive activation of p70s6k by endogenous and heterologously expressed Gi-coupled receptors, *J. Biol. Chem.* **271** (1996) 8537–8540.
257. H.B. Jefferies, C. Reinhard, S.C. Kozma and G. Thomas, Rapamycin selectively represses translation of the "polypyrimidine tract" mRNA family, *Proc. Natl. Acad. Sci. USA* **91** (1994) 4441–4445.
258. M.M. Chou and J. Blenis, The 70 kDa S6 kinase complexes with and is activated by the Rho family G proteins Cdc42 and Rac1, *Cell* **85** (1996) 573–583.
259. P.J. Parker and J. Murray-Rust, PKC at a glance, *J. Cell Sci.* **117** (2004) 131–132.
260. J.A. Le Good, W.H. Ziegler, D.B. Parekh, D.R. Alessi, P. Cohen and P.J. Parker, Protein kinase C isotypes controlled by phosphoinositide 3-kinase through the protein kinase PDK1, *Science* **281** (1998) 2042–2045.
261. K. Akimoto, R. Takahashi, S. Moriya, N. Nishioka, J. Takayanagi, K. Kimura, Y. Fukui, S. Osada, K. Mizuno, S. Hirai, A. Kazlauskas and S. Ohno, EGF or PDGF receptors activate atypical PKClambda through phosphatidylinositol 3-kinase, *EMBO J.* **15** (1996) 788–798.
262. K. Kotani, W. Ogawa, M. Matsumoto, T. Kitamura, H. Sakaue, Y. Hino, K. Miyake, W. Sano, K. Akimoto, S. Ohno and M. Kasuga, Requirement of atypical protein kinase clambda for insulin stimulation of glucose uptake but not for Akt activation in 3T3-L1 adipocytes, *Mol. Cell. Biol.* **18** (1998) 6971–6982.
263. E.M. Dutil, A. Toker and A.C. Newton, Regulation of conventional protein kinase C isozymes by phosphoinositide-dependent kinase 1 (PDK-1), *Curr. Biol.* **8** (1998) 1366–1375.
264. S.S. Taylor, J. Yang, J. Wu, N.M. Haste, E. Radzio-Andzelm and G. Anand, PKA: a portrait of protein kinase dynamics, *Biochim. Biophys. Acta* **1697** (2004) 259–269.

265. J.J. Michel and J.D. Scott, AKAP mediated signal transduction, *Annu. Rev. Pharmacol. Toxicol.* **42** (2002) 235–257.
266. M.J. Moore, J.R. Kanter, K.C. Jones and S.S. Taylor, Phosphorylation of the catalytic submit of protein Kinase A: autophosphorylation versus phosphorylation by phosphoinositide-dependent kinase-1, *J. Biol. Chem.* **277** (2002) 47878–47884.
267. J.R. Bayascas and D.R. Alessi, Regulation of Akt/PKB Ser473 phosphorylation, *Mol. Cell* **18** (2005) 143–145.
268. A. Toker and A.C. Newton, Cellular signaling: pivoting around PDK-1, *Cell* **103** (2000) 185–188.
269. A.A. Troussard, N.M. Mawji, C. Ong, A. Mui, R. Arnaud and S. Dedhar, Conditional knock-out of integrin-linked kinase demonstrates an essential role in protein kinase B/Akt activation, *J. Biol. Chem.* **278** (2003) 22374–22378.
270. K. Hara, Y. Maruki, X. Long, K. Yoshino, N. Oshiro, S. Hidayat, C. Tokunaga, J. Avruch and K. Yonezawa, Raptor, a binding partner of target of rapamycin (TOR), mediates TOR action, *Cell* **110** (2002) 177–189.
271. D.H. Kim, D.D. Sarbassov, S.M. Ali, J.E. King, R.R. Latek, H. Erdjument-Bromage, P. Tempst and D.M. Sabatini, mTOR interacts with raptor to form a nutrient-sensitive complex that signals to the cell growth machinery, *Cell* **110** (2002) 163–175.
272. D.D. Sarbassov, D.A. Guertin, S.M. Ali and D.M. Sabatini, Phosphorylation and regulation of Akt/PKB by the rictor-mTOR complex, *Science* **307** (2005) 1098–1101.
273. W. Zhang and L. Shi, Evolution of the PPM-family protein phosphatases in Streptomyces: duplication of catalytic domain and lateral recruitment of additional sensory domains, *Microbiology* **150** (2004) 4189–4197.
274. W.S. Chen, P.Z. Xu, K. Gottlob, M.L. Chen, K. Sokol, T. Shiyanova, I. Roninson, W. Weng, R. Suzuki, K. Tobe, T. Kadowaki and N. Hay, Growth retardation and increased apoptosis in mice with homozygous disruption of the Akt1 gene, *Genes Dev.* **15** (2001) 2203–2208.
275. Z.Z. Yang, O. Tschopp, M. Hemmings-Mieszczak, J. Feng, D. Brodbeck, E. Perentes and B.A. Hemmings, Protein kinase B alpha/Akt1 regulates placental development and fetal growth, *J. Biol. Chem.* **278** (2003) 32124–32131.
276. J. Chen, S. De, D.S. Damron, W.S. Chen, N. Hay and T.V. Byzova, Impaired platelet responses to thrombin and collagen in AKT-1-deficient mice, *Blood* **104** (2004) 1703–1710.

277. H. Cho, J. Mu, J.K. Kim, J.L. Thorvaldsen, Q. Chu, E.B. Crenshaw, III, K.H. Kaestner, M.S. Bartolomei, G.I. Shulman and M.J. Birnbaum, Insulin resistance and a diabetes mellitus-like syndrome in mice lacking the protein kinase Akt2 (PKB beta), *Science* **292** (2001) 1728–1731.
278. D. Kim, H.C. Dan, S. Park, L. Yang, Q. Liu, S. Kaneko, J. Ning, L. He, H. Yang, M. Sun, S.V. Nicosia and J.Q. Cheng, AKT/PKB signaling mechanisms in cancer and chemoresistance, *Front Biosci.* **10** (2005) 975–987.
279. T.F. Franke, D.R. Kaplan and L.C. Cantley, PI3K: downstream AKTion blocks apoptosis, *Cell* **88** (1997) 435–437.
280. D.T. Chao and S.J. Korsmeyer, BCL-2 family: regulators of cell death, *Annu. Rev. Immunol.* **16** (1998) 395–419.
281. X. Fang, S. Yu, A. Eder, M. Mao, R.C. Bast, Jr., D. Boyd and G.B. Mills, Regulation of BAD phosphorylation at serine 112 by the Ras-mitogen-activated protein kinase pathway, *Oncogene* **18** (1999) 6635–6640.
282. P.L. del, M. Gonzalez-Garcia, C. Page, R. Herrera and G. Nunez, Interleukin-3-induced phosphorylation of BAD through the protein kinase Akt, *Science* **278** (1997) 687–689.
283. S.R. Datta, H. Dudek, X. Tao, S. Masters, H. Fu, Y. Gotoh and M.E. Greenberg, Akt phosphorylation of BAD couples survival signals to the cell intrinsic death machinery, *Cell* **91** (1997) 231–241.
284. M.H. Cardone, N. Roy, H.R. Stennicke, G.S. Salvesen, T.F. Franke, E. Stanbridge, S. Frisch and J.C. Reed, Regulation of cell death protease caspase-9 by phosphorylation, *Science* **282** (1998) 1318–1321.
285. I. Shin, F.M. Yakes, F. Rojo, N.Y. Shin, A.V. Bakin, J. Baselga and C.L. Arteaga, PKB/Akt mediates cell-cycle progression by phosphorylation of p27(Kip1) at threonine 157 and modulation of its cellular localization, *Nat. Med.* **8** (2002) 1145–1152.
286. G. Viglietto, M.L. Motti, P. Bruni, R.M. Melillo, A. D'Alessio, D. Califano, F. Vinci, G. Chiappetta, P. Tsichlis, A. Bellacosa, A. Fusco and M. Santoro, Cytoplasmic relocalization and inhibition of the cyclin-dependent kinase inhibitor p27(Kip1) by PKB/Akt-mediated phosphorylation in breast cancer, *Nat. Med.* **8** (2002) 1136–1144.
287. B.P. Zhou, Y. Liao, W. Xia, B. Spohn, M.H. Lee and M.C. Hung, Cytoplasmic localization of p21Cip1/WAF1 by Akt-induced phosphorylation in HER-2/neu-overexpressing cells, *Nat. Cell Biol.* **3** (2001) 245–252.
288. L.D. Mayo and D.B. Donner, A phosphatidylinositol 3-kinase/Akt pathway promotes translocation of Mdm2 from the cytoplasm to the nucleus, *Proc. Natl. Acad. Sci. USA* **98** (2001) 11598–11603.

289. B.P. Zhou, Y. Liao, W. Xia, Y. Zou, B. Spohn and M.C. Hung, HER-2/neu induces p53 ubiquitination via Akt-mediated MDM2 phosphorylation, *Nat. Cell Biol.* **3** (2001) 973–982.
290. P.J. Coffer, B.M. Burgering, Forkhead-box transcription factors and their role in the immune system, *Nat. Rev. Immunol.* **4** (2004) 889–899.
291. G.J. Kops, N.D. de Ruiter, A.M. Vries-Smits, D.R. Powell, J.L. Bos and B.M. Burgering, Direct control of the Forkhead transcription factor AFX by protein kinase B, *Nature* **398** (1999) 630–634.
292. L.P. Van Der Heide, M.F. Hoekman and M.P. Smidt, The ins and outs of FoxO shuttling: mechanisms of FoxO translocation and transcriptional regulation, *Biochem. J.* **380** (2004) 297–309.
293. W.H. Biggs, III, J. Meisenhelder, T. Hunter, W.K. Cavenee and K.C. Arden, Protein kinase B/Akt-mediated phosphorylation promotes nuclear exclusion of the winged helix transcription factor FKHR1, *Proc. Natl. Acad. Sci. USA* **96** (1999) 7421–7426.
294. H. Takaishi, H. Konishi, H. Matsuzaki, Y. Ono, Y. Shirai, N. Saito, T. Kitamura, W. Ogawa, M. Kasuga, U. Kikkawa and Y. Nishizuka, Regulation of nuclear translocation of forkhead transcription factor AFX by protein kinase B, *Proc. Natl. Acad. Sci. USA* **96** (1999) 11836–11841.
295. A. Brunet, A. Bonni, M.J. Zigmond, M.Z. Lin, P. Juo, L.S. Hu, M.J. Anderson, K.C. Arden, J. Blenis and M.E. Greenberg, Akt promotes cell survival by phosphorylating and inhibiting a Forkhead transcription factor, *Cell* **96** (1999) 857–868.
296. A. Brunet, J. Park, H. Tran, L.S. Hu, B.A. Hemmings and M.E. Greenberg, Protein kinase SGK mediates survival signals by phosphorylating the forkhead transcription factor FKHRL1 (FOXO3a), *Mol. Cell. Biol.* **21** (2001) 952–965.
297. O.N. Ozes, L.D. Mayo, J.A. Gustin, S.R. Pfeffer, L.M. Pfeffer and D.B. Donner, NF-kappaB activation by tumour necrosis factor requires the Akt serine-threonine kinase, *Nature* **401** (1999) 82–85.
298. L.P. Kane, V.S. Shapiro, D. Stokoe and A. Weiss, Induction of NF-kappaB by the Akt/PKB kinase, *Curr. Biol.* **9** (1999) 601–604.
299. A. Lauder, A. Castellanos and K. Weston, c-Myb transcription is activated by protein kinase B (PKB) following interleukin 2 stimulation of Tcells and is required for PKB-mediated protection from apoptosis, *Mol. Cell. Biol.* **21** (2001) 5797–5805.
300. K. Du and M. Montminy, CREB is a regulatory target for the protein kinase Akt/PKB, *J. Biol. Chem.* **273** (1998) 32377–32379.

301. S. Pugazhenthi, A. Nesterova, C. Sable, K.A. Heidenreich, L.M. Boxer, L.E. Heasley and J.E. Reusch, Akt/protein kinase B up-regulates Bcl-2 expression through cAMP-response element-binding protein, *J. Biol. Chem.* **275** (2000) 10761–10766.
302. J.M. Wang, J.R. Chao, W. Chen, M.L. Kuo, J.J. Yen and H.F. Yang-Yen, The antiapoptotic gene mcl-1 is up-regulated by the phosphatidylinositol 3-kinase/Akt signaling pathway through a transcription factor complex containing CREB, *Mol. Cell. Biol.* **19** (1999) 6195–6206.
303. N. Nakamura, S. Ramaswamy, F. Vazquez, S. Signoretti, M. Loda and W.R. Sellers, Forkhead transcription factors are critical effectors of cell death and cell cycle arrest downstream of PTEN, *Mol. Cell. Biol.* **20** (2000) 8969–8982.
304. I. Ciechomska, B. Pyrzynska, P. Kazmierczak and B. Kaminska, Inhibition of Akt kinase signalling and activation of Forkhead are indispensable for upregulation of fast expression in apoptosis of glioma cells, *Oncogene* **22** (2003) 7617–7627.
305. C.M. Cahill, G. Tzivion, N. Nasrin, S. Ogg, J. Dore, G. Ruvkun and M. Alexander-Bridges, Phosphatidylinositol 3-kinase signaling inhibits DAF-16 DNA binding and function via 14-3-3-dependent and 14-3 3-independent pathways, *J. Biol. Chem.* **276** (2001) 13402–13410.
306. B.B. Aggarwal, Nuclear factor-kappaB: the enemy within, *Cancer Cell* **6** (2004) 203–208.
307. A. Khwaja, Akt is more than just a Bad kinase, *Nature* **401** (1999) 33–34.
308. B.C. Kim, M.N. Lee, J.Y. Kim, S.S. Lee, J.D. Chang, S.S. Kim, S.Y. Lee and J.H. Kim, Roles of phosphatidylinositol 3-kinase and Rac in the nuclear signaling by tumor necrosis factor-alpha in rat-2 fibroblasts, *J. Biol. Chem.* **274** (1999) 24372–24377.
309. Y. Wang, M. Falasca, J. Schlessinger, S. Malstrom, P. Tsichlis, J. Settleman, W. Hu, B. Lim and R. Prywes, Activation of the c-fos serum response element by phosphatidyl inositol 3-kinase and rho pathways in HeLa cells, *Cell Growth Differ.* **9** (1998) 513–522.
310. M.D. ConKright and M. Montminy, CREB: the unindicted cancer co-conspirator, *Trends Cell Biol.* **15** (2005) 457–459.
311. B. Mayr and M. Montminy, Transcriptional regulation by the phosphorylation-dependent factor CREB, *Nat. Rev. Mol. Cell Biol.* **2** (2001) 599–609.
312. P.J. Coffer, J. Jin and J.R. Woodgett, Protein kinase B (c-Akt): a multifunctional mediator of phosphatidylinositol 3-kinase activation, *Biochem. J.* **335**(Pt 1) (1998) 1–13.

313. M. Delcommenne, C. Tan, V. Gray, L. Rue, J. Woodgett and S. Dedhar, Phosphoinositide-3-OH kinase-dependent regulation of glycogen synthase kinase 3 and protein kinase B/AKT by the integrin-linked kinase, *Proc. Natl. Acad. Sci. USA* **95** (1998) 11211–11216.
314. G.I. Welsh, C.M. Miller, A.J. Loughlin, N.T. Price and C.G. Proud, Regulation of eukaryotic initiation factor eIF2B: glycogen synthase kinase-3 phosphorylates a conserved serine which undergoes dephosphorylation in response to insulin, *FEBS Lett.* **421** (1998) 125–130.
315. C.G. Proud, eIF2 and the control of cell physiology, *Semin. Cell Dev. Biol.* **16** (2005) 3–12.
316. X. Wang, M. Janmaat, A. Beugnet, F.E. Paulin and C.G. Proud, Evidence that the dephosphorylation of Ser(535) in the epsilon-subunit of eukaryotic initiation factor (eIF) 2B is insufficient for the activation of eIF2B by insulin, *Biochem. J.* **367** (2002) 475–481.
317. J. Deprez, D. Vertommen, D.R. Alessi, L. Hue and M.H. Rider, Phosphorylation and activation of heart 6-phosphofructo-2-kinase by protein kinase B and other protein kinases of the insulin signaling cascades, *J. Biol. Chem.* **272** (1997) 17269–17275.
318. M.H. Rider, L. Bertrand, D. Vertommen, P.A. Michels, G.G. Rousseau and L. Hue, 6-phosphofructo-2-kinase/fructose-2,6-bisphosphatase: head-to-head with a bifunctional enzyme that controls glycolysis, *Biochem. J.* **381** (2004) 561–579.
319. A.C. Gingras, B. Raught and N. Sonenberg, Regulation of translation initiation by FRAP/mTOR, *Genes Dev.* **15** (2001) 807–826.
320. A.F. Shamji, P. Nghiem and S.L. Schreiber, Integration of growth factor and nutrient signaling: implications for cancer biology, *Mol. Cell* **12** (2003) 271–280.
321. D.C. Fingar and J. Blenis, Target of rapamycin (TOR): an integrator of nutrient and growth factor signals and coordinator of cell growth and cell cycle progression, *Oncogene* **23** (2004) 3151–3171.
322. K. Inoki, Y. Li, T. Zhu, J. Wu and K.L. Guan, TSC2 is phosphorylated and inhibited by Akt and suppresses mTOR signalling, *Nat. Cell Biol.* **4** (2002) 648–657.
323. C.J. Potter, L.G. Pedraza and T. Xu, Akt regulates growth by directly phosphorylating Tsc2, *Nat. Cell Biol.* **4** (2002) 658–665.
324. X. Gao and D. Pan, TSC1 and TSC2 tumor suppressors antagonize insulin signaling in cell growth, *Genes Dev.* **15** (2001) 1383–1392.
325. D.J. Kwiatkowski, Rhebbing up mTOR: new insights on TSC1 and TSC2, and the pathogenesis of tuberous sclerosis, *Cancer Biol. Ther.* **2** (2003) 471–476.

326. I.K. Mellinghoff and C.L. Sawyers, TORward AKTually useful mouse models, *Nat. Med.* **10** (2004) 579–580.
327. L.J. Berg, L.D. Finkelstein, J.A. Lucas and P.L. Schwartzberg, TEC family kinases in T lymphocyte development and function, *Annu. Rev. Immunol.* **23** (2005) 549–600.
328. Z. Li, M.I. Wahl, A. Eguinoa, L.R. Stephens, P.T. Hawkins and O.N. Witte, Phosphatidylinositol 3-kinase-gamma activates Bruton's tyrosine kinase in concert with Src family kinases, *Proc. Natl. Acad. Sci. USA* **94** (1997) 13820–13825.
329. A. August, A. Sadra, B. Dupont and H. Hanafusa, Src-induced activation of inducible T cell kinase (ITK) requires phosphatidylinositol 3-kinase activity and the Pleckstrin homology domain of inducible T cell kinase, *Proc. Natl. Acad. Sci. USA* **94** (1997) 11227–11232.
330. M.E. Conley, A. Broides, V. Hernandez-Trujillo, V. Howard, H. Kanegane, T. Miyawaki and S.A. Shurtleff, Genetic analysis of patients with defects in early B-cell development, *Immunol. Rev.* **203** (2005) 216–234.
331. J.M. Lindvall, K.E. Blomberg, J. Valiaho, L. Vargas, J.E. Heinonen, A. Berglof, A.J. Mohamed, B.F. Nore, M. Vihinen and C.I. Smith, Bruton's tyrosine kinase: cell biology, sequence conservation, mutation spectrum, siRNA modifications, and expression profiling, *Immunol. Rev.* **203** (2005) 200–215.
332. D.J. Rawlings, D.C. Saffran, S. Tsukada, D.A. Largaespada, J.C. Grimaldi, L. Cohen, R.N. Mohr, J.F. Bazan, M. Howard, N.G. Copeland, *et al.*, Mutation of unique region of Bruton's tyrosine kinase in immunodeficient XID mice, *Science* **261** (1993) 358–361.
333. K. Salim, M.J. Bottomley, E. Querfurth, M.J. Zvelebil, I. Gout, R. Scaife, R.L. Margolis, R. Gigg, C.I. Smith and P.C. Driscoll, Distinct specificity in the recognition of phosphoinositides by the pleckstrin homology domains of dynamin and Bruton's tyrosine kinase, *EMBO J.* **15** (1996) 6241–6250.
334. H. Suzuki, Y. Terauchi, M. Fujiwara, S. Aizawa, Y. Yazaki, T. Kadowaki and S. Koyasu, Xid-like immunodeficiency in mice with disruption of the p85alpha subunit of phosphoinositide 3-kinase, *Science* **283** (1999) 390–392.
335. A.M. Buhl and J.C. Cambier, Phosphorylation of CD19 Y484 and Y515, and linked activation of phosphatidylinositol 3-kinase, are required for B cell antigen receptor-mediated activation of Bruton's tyrosine kinase, *J. Immunol.* **162** (1999) 4438–4446.
336. S.G. Rhee, Regulation of phosphoinositide-specific phospholipase C, *Annu. Rev. Biochem.* **70** (2001) 281–312.
337. M.J. Berridge, Unlocking the secrets of cell signaling, *Annu. Rev. Physiol* **67** (2005) 1–21.

338. R. Bobe, J.I. Wilde, P. Maschberger, K. Venkateswarlu, P.J. Cullen, W. Siess and S.P. Watson, Phosphatidylinositol 3-kinase-dependent translocation of phospholipase Cgamma2 in mouse megakaryocytes is independent of Bruton tyrosine kinase translocation, *Blood* **97** (2001) 678–684.
339. A. Wittinghofer, Signal transduction via Ras, *Biol. Chem.* **379** (1998) 933–937.
340. D.K. Worthylake, K.L. Rossman and J. Sondek, Crystal structure of the DH/PH fragment of Dbs without bound GTPase, *Structure (Camb.)* **12** (2004) 1078–1086.
341. H.C. Welch, W.J. Coadwell, L.R. Stephens and P.T. Hawkins, Phosphoinositide 3-kinase-dependent activation of Rac, *FEBS Lett.* **546** (2003) 93–97.
342. L.E. Rameh and L.C. Cantley, The role of phosphoinositide 3-kinase lipid products in cell function, *J. Biol. Chem.* **274** (1999) 8347–8350.
343. J. Han, K. Luby-Phelps, B. Das, X. Shu, Y. Xia, R.D. Mosteller, U.M. Krishna, J.R. Falck, M.A. White and D. Broek, Role of substrates and products of PI 3-kinase in regulating activation of Rac-related guanosine triphosphatases by Vav, *Science* **279** (1998) 558–560.
344. K. Venkateswarlu, P.B. Oatey, J.M. Tavare and P.J. Cullen, Insulin-dependent translocation of ARNO to the plasma membrane of adipocytes requires phosphatidylinositol 3-kinase, *Curr. Biol.* **8** (1998) 463–466.
345. P.J. Lockyer, S. Wennstrom, S. Kupzig, K. Venkateswarlu, J. Downward and P.J. Cullen, Identification of the ras GTPase-activating protein GAP1(m) as a phosphatidylinositol-3,4,5-trisphosphate-binding protein in vivo, *Curr. Biol.* **9** (1999) 265–268.
346. K. Venkateswarlu, P.B. Oatey, J.M. Tavare, T.R. Jackson and P.J. Cullen, Identification of centaurin-alpha1 as a potential in vivo phosphatidylinositol 3,4,5-trisphosphate-binding protein that is functionally homologous to the yeast ADP-ribosylation factor (ARF) GTPase-activating protein, Gcs1, *Biochem. J.* **340**(Pt 2) (1999) 359–363.
347. K. Kotani, K. Yonezawa, K. Hara, H. Ueda, Y. Kitamura, H. Sakaue, A. Ando, A. Chavanieu, B. Calas, F. Grigorescu, *et al.*, Involvement of phosphoinositide 3-kinase in insulin- or IGF-1-induced membrane ruffling, *EMBO J.* **13** (1994) 2313–2321.
348. K. Hara, K. Yonezawa, H. Sakaue, A. Ando, K. Kotani, T. Kitamura, Y. Kitamura, H. Ueda, L. Stephens, T.R. Jackson, P.T. Hawkins, R. Dhand, A.E. Clark, G.D. Holman, M.D. Waterfield and M. Kasuga, 1-Phosphatidylinositol 3-kinase activity is required for insulin-stimulated glucose transport but not for RAS activation in CHO cells, *Proc. Natl. Acad. Sci. USA* **91** (1994) 7415–7419.

349. S. Roche, M. Koegl, M.V. Barone, M.F. Roussel and S.A. Courtneidge, DNA synthesis induced by some but not all growth factors requires Src family protein tyrosine kinases, *Mol. Cell. Biol.* **15** (1995) 1102–1109.
350. A. Yart, S. Roche, R. Wetzker, M. Laffargue, N. Tonks, P. Mayeux, H. Chap and P. Raynal, A function for phosphoinositide 3-kinase beta lipid products in coupling beta gamma to Ras activation in response to lysophosphatidic acid, *J. Biol. Chem.* **277** (2002) 21167–21178.
351. B. Vanhaesebroeck, G.E. Jones, W.E. Allen, D. Zicha, R. Hooshmand-Rad, C. Sawyer, C. Wells, M.D. Waterfield and A.J. Ridley, Distinct PI(3)Ks mediate mitogenic signalling and cell migration in macrophages, *Nat. Cell Biol.* **1** (1999) 69–71.
352. N. Schutze, siRNA technology, *Mol. Cell Endocrinol.* **213** (2004) 115–119.
353. D.A. Fruman, F. Mauvais-Jarvis, D.A. Pollard, C.M. Yballe, D. Brazil, R.T. Bronson, C.R. Kahn and L.C. Cantley, Hypoglycaemia, liver necrosis and perinatal death in mice lacking all isoforms of phosphoinositide 3-kinase p85 alpha, *Nat. Genet.* **26** (2000) 379–382.
354. K. Ueki, C.M. Yballe, S.M. Brachmann, D. Vicent, J.M. Watt, C.R. Kahn and L.C. Cantley, Increased insulin sensitivity in mice lacking p85beta subunit of phosphoinositide 3-kinase, *Proc. Natl. Acad. Sci. USA* **99** (2002) 419–424.
355. S.M. Brachmann, C.M. Yballe, M. Innocenti, J.A. Deane, D.A. Fruman, S.M. Thomas and L.C. Cantley, Role of phosphoinositide 3-kinase regulatory isoforms in development and actin rearrangement, *Mol. Cell. Biol.* **25** (2005) 2593–2606.
356. D.A. Fruman, S.B. Snapper, C.M. Yballe, L. Davidson, J.Y. Yu, F.W. Alt and L.C. Cantley, Impaired B cell development and proliferation in absence of phosphoinositide 3-kinase p85alpha, *Science* **283** (1999) 393–397.
357. L. Bi, I. Okabe, D.J. Bernard, A. Wynshaw-Boris and R.L. Nussbaum, Proliferative defect and embryonic lethality in mice homozygous for a deletion in the p110alpha subunit of phosphoinositide 3-kinase, *J. Biol. Chem.* **274** (1999) 10963–10968.
358. L. Bi, I. Okabe, D.J. Bernard and R.L. Nussbaum, Early embryonic lethality in mice deficient in the p110beta catalytic subunit of PI 3-kinase, *Mamm. Genome* **13** (2002) 169–172.
359. S.M. Brachmann, K. Ueki, J.A. Engelman, R.C. Kahn and L.C. Cantley, Phosphoinositide 3-kinase catalytic subunit deletion and regulatory subunit deletion have opposite effects on insulin sensitivity in mice, *Mol. Cell. Biol.* **25** (2005) 1596–1607.

360. J.M. Lizcano, S. Alrubaie, A. Kieloch, M. Deak, S.J. Leevers and D.R. Alessi, Insulin-induced Drosophila S6 kinase activation requires phosphoinositide 3-kinase and protein kinase B, *Biochem. J.* **374** (2003) 297–306.
361. M. Miron, P. Lasko and N. Sonenberg, Signaling from Akt to FRAP/TOR targets both 4E-BP and S6K in Drosophila melanogaster, *Mol. Cell. Biol.* **23** (2003) 9117–9126.
362. F. Czanderna, A. Santel, M. Hinz, M. Fechtner, B. Durieux, G. Fisch, F. Leenders, W. Arnold, K. Giese, A. Klippel and J. Kaufmann, Inducible shRNA expression for application in a prostate cancer mouse model, *Nucleic Acids Res.* **31** (2003) e127.
363. S.T. Jou, N. Carpino, Y. Takahashi, R. Piekorz, J.R. Chao, N. Carpino, D. Wang and J.N. Ihle, Essential, nonredundant role for the phosphoinositide 3-kinase p110delta in signaling by the B-cell receptor complex, *Mol. Cell. Biol.* **22** (2002) 8580–8591.
364. E. Clayton, G. Bardi, S.E. Bell, D. Chantry, C.P. Downes, A. Gray, L.A. Humphries, D. Rawlings, H. Reynolds, E. Vigorito and M. Turner, A crucial role for the p110delta subunit of phosphatidylinositol 3-kinase in B cell development and activation, *J. Exp. Med.* **196** (2002) 753–763.
365. K. Okkenhaug, A. Bilancio, G. Farjot, H. Priddle, S. Sancho, E. Peskett, W. Pearce, S.E. Meek, A. Salpekar, M.D. Waterfield, A.J. Smith and B. Vanhaesebroeck, Impaired B and T cell antigen receptor signaling in p110delta PI 3-kinase mutant mice, *Science* **297** (2002) 1031–1034.
366. M. Laffargue, R. Calvez, P. Finan, A. Trifilieff, M. Barbier, F. Altruda, E. Hirsch and M.P. Wymann, Phosphoinositide 3-kinase gamma is an essential amplifier of mast cell function, *Immunity* **16** (2002) 441–451.
367. E. Hirsch, V.L. Katanaev, C. Garlanda, O. Azzolino, L. Pirola, L. Silengo, S. Sozzani, A. Mantovani, F. Altruda and M.P. Wymann, Central role for G protein-coupled phosphoinositide 3-kinase gamma in inflammation, *Science* **287** (2000) 1049–1053.
368. Z. Li, H. Jiang, W. Xie, Z. Zhang, A.V. Smrcka and D. Wu, Roles of PLC-beta2 and -beta3 and PI3Kgamma in chemoattractant-mediated signal transduction, *Science* **287** (2000) 1046–1049.
369. T. Sasaki, J. Irie-Sasaki, R. Jones, A. Oliveira-dos-Santos, W. Stanford, B. Bolon, A. Wakeham, A. Itie, D. Bouchard, I. Kozieradzki, N. Joza, T. Mak, P. Ohashi, A. Suzuki and J. Penninger, Function of PI3K gamma in thymocyte development, T cell activation, and neutrophil migration, *Science* **287** (2000) 1040–1046.
370. D.A. Fruman, Towards an understanding of isoform specificity in phosphoinositide 3-kinase signalling in lymphocytes, *Biochem. Soc. Trans.* **32** (2004) 315–319.

371. K. Ali, A. Bilancio, M. Thomas, W. Pearce, A.M. Gilfillan, C. Tkaczyk, N. Kuehn, A. Gray, J. Giddings, E. Peskett, R. Fox, I. Bruce, C. Walker, C. Sawyer, K. Okkenhaug, P. Finan and B. Vanhaesebroeck, Essential role for the p110delta phosphoinositide 3-kinase in the allergic response, *Nature* **431** (2004) 1007–1011.
372. G.E. Jones, E. Prigmore, R. Calvez, C. Hogan, G.A. Dunn, E. Hirsch, M.P. Wymann and A.J. Ridley, Requirement for PI3-kinase gamma in macrophage migration to MCP-1 and CSF-1, *Exp. Cell Res.* **290** (2003) 120–131.
373. M. Hannigan, L. Zhan, Z. Li, Y. Ai, D. Wu and C.K. Huang, Neutrophils lacking phosphoinositide 3-kinase gamma show loss of directionality during N-formyl-Met-Leu-Phe-induced chemotaxis, *Proc. Natl. Acad. Sci. USA* **99** (2002) 3603–3608.
374. P.E. Row, B.J. Reaves, J. Domin, J.P. Luzio and H.W. Davidson, Overexpression of a rat kinase-deficient phosphoinositide 3-kinase, Vps34p, inhibits cathepsin D maturation, *Biochem. J.* **353** (2001) 655–661.
375. A. Simonsen, R. Lippe, S. Christoforidis, J.M. Gaullier, A. Brech, J. Callaghan, B.H. Toh, C. Murphy, M. Zerial and H. Stenmark, EEA1 links PI(3)K function to Rab5 regulation of endosome fusion, *Nature* **394** (1998) 494–498.
376. S.F. Hausdorff, D.C. Fingar, K. Morioka, L.A. Garza, E.L. Whiteman, S.A. Summers and M.J. Birnbaum, Identification of wortmannin-sensitive targets in 3T3 L1 adipocytes. Dissociationo of insulin-stimulated glucose uptake and GLUT4 translocation, *J. Biol. Chem.* **274** (1999) 24677–24684.
377. D. Malide and S.W. Cushman, Morphological effects of wortmannin on the endosomal system and GLUT4-containing compartments in rat adipose cells, *J. Cell Sci.* **110** (Pt 22) (1997) 2795–2806.
378. S. Volinia, R. Dhand, B. Vanhaesebroeck, L.K. MacDougall, R. Stein, M.J. Zvelebil, J. Domin, C. Panaretou and M.D. Waterfield, A human phosphatidylinositol 3-kinase complex related to the yeast Vps34p-Vps15p protein sorting system, *EMBO J.* **14** (1995) 3339–3348.
379. U. Siddhanta, J. McIlroy, A. Shah, Y. Zhang and J.M. Backer, Distinct roles for the p110alpha and hVPS34 phosphatidylinositol 3′-kinases in vesicular trafficking, regulation of the actin cytoskeleton, and mitogenesis, *J. Cell Biol.* **143** (1998) 1647–1659.
380. C. Chaussade, L. Pirola, S. Bonnafous, F. Blondeau, S. Brenz-Verca, H. Tronchere, F. Portis, S. Rusconi, B. Payrastre, J. Laporte and O.E. Van, Expression of myotubularin by an adenoviral vector demonstrates its function as a phosphatidylinositol 3-phosphate [PtdIns(3)P] phosphatase in muscle cell lines: involvement of PtdIns(3)P in insulin-stimulated glucose transport, *Mol. Endocrinol.* **17** (2003) 2448–2460.

381. S.L. Macaulay, V. Stoichevska, J. Grusovin, K.H. Gough, L.A. Castelli and C.W. Ward, Insulin stimulates movement of sorting nexin 9 between cellular compartments: a putative role mediating cell surface receptor expression and insulin action, *Biochem. J.* **376** (2003) 123–134.
382. D.J. Klionsky, The molecular machinery of autophagy: unanswered questions, *J. Cell Sci.* **118** (2005) 7–18.
383. D. Gozuacik and A. Kimchi, Autophagy as a cell death and tumor suppressor mechanism, *Oncogene* **23** (2004) 2891–2906.
384. X. Zeng, J.H. Overmeyer and W.A. Maltese, Functional specificity of the mammalian Beclin-Vps34 PI 3-kinase complex in macroautophagy versus endocytosis and lysosomal enzyme trafficking, *J. Cell Sci.* **119** (2006) 259–270.
385. A. Tassa, M.P. Roux, D. Attaix and D.M. Bechet, Class III phosphoinositide 3-kinase-Beclin 1 complex mediates the amino acid dependent regulation of autophagy in C2C12 myotubes, *Biochem. J.* **376** (2003) 577–586.
386. A. Kihara, Y. Kabeya, Y. Ohsumi and T. Yoshimori, Beclin-phosphatidylinositol 3-kinase complex functions at the trans-Golgi network, *EMBO Rep.* **2** (2001) 330–335.
387. E.L. Eskelinen, A.R. Prescott, J. Cooper, S.M. Brachmann, L. Wang, X. Tang, J.M. Backer and J.M. Lucocq, Inhibition of autophagy in mitotic animal cells, *Traffic* **3** (2002) 878–893.
388. T. Maffucci, A. Brancaccio, E. Piccolo, R.C. Stein and M. Falasca, Insulin induces phosphatidylinositol-3-phosphate formation through TC10 activation, *EMBO J.* **22** (2003) 4178–4189.
389. I. Gaidarov, Y. Zhao and J.H. Keen, Individual phosphoinositide 3-kinase C2 alpha domain activities independently regulate clathrin function, *J. Biol. Chem.* **280** (2005) 40766–40772.
390. J.L. Martys, C. Wjasous, D.M. Gangi, M.C. Kielian, T.E. McGraw, J.M. Backer, Wortmannin-sensitive trafficking pathways in Chinese hamster ovary cells: differential effects on endocytosis and lysosomal sorting, *J. Biol. Chem.* **271** (1996) 10953–10962.
391. M.D. Chamberlain, T.R. Berry, M.C. Pastor and D.H. Anderson, The p85alpha subunit of phosphatidylinositol 3′-kinase binds to and stimulates the GTPase activity of Rab proteins, *J. Biol. Chem.* **279** (2004) 48607–48614.
392. S. Christoforidis, M. Miaczynska, K. Ashman, M. Wilm, L. Zhao, S. Yip, M. Waterfield, J. Backer and M. Zerial, Phosphatidylinositol-3-OH kinases are Rab5 effectors, *Nat. Cell Biol.* **1** (1999) 249–252.
393. A.T. Jones, D.J. Spiro, T. Kirchhausen, P. Melancon and M. Wessling-Resnick, Studies on the inhibition of endosome fusion by GTPgammaS-bound ARF, *J. Cell Sci.* **112**(Pt 20) (1999) 3477–3485.

394. M. Joly, A. Kazlauskas, F.S. Fay and S. Corvera, Disruption of PDGF receptor trafficking by mutation of its PI-3 kinase binding sites, *Science* **263** (1994) 684–687.
395. M. Joly, A. Kazlauskas and S. Corvera, Phosphatidylinositol 3-kinase activity is required at a postendocytic step in platelet-derived growth factor receptor trafficking, *J. Biol. Chem.* **270** (1995) 13225–13230.
396. H. Shpetner, M. Joly, D. Hartley and S. Corvera, Potential sites of PI-3 kinase function in the endocytic pathway revealed by the PI-3 kinase inhibitor, wortmannin, *J. Cell Biol.* **132** (1996) 595–605.
397. S. Ogg and G. Ruvkun, The C. elegans PTEN homolog, DAF-18, acts in the insulin receptor-like metabolic signaling pathway, *Mol. Cell* **2** (1998) 887–893.
398. J.Z. Morris, H.A. Tissenbaum and G. Ruvkun, A phosphinositol 3-OH kinase family member regulating longevity and diapause in *Caenorhabditis elegans, Nature* **382** (1996) 536–539.
399. H. Stocker, M. Andjelkovic, S. Oldham, M. Laffargue, M.P. Wymann, B.A. Hemmings and E. Hafen, Living with lethal PIP3 levels: viability of flies lacking PTEN restored by a PH domain mutation in Akt/PKB, *Science* **295** (2002) 2088–2091.
400. M. Tresini, M. Mawal-Dewan, V.J. Cristofalo and C. Sell, A phosphatidylinositol 3-kinase inhibitor induces a senescent-like growth arrest in human diploid fibroblasts, *Cancer Res.* **58** (1998) 1–4.
401. S. Roche, M. Koegl and S.A. Courtneidge, The phosphatidylinositol 3-kinase alpha is required for DNA synthesis induced by some, but not all, growth factors, *Proc. Natl. Acad. Sci. USA* **91** (1994) 9185–9189.
402. M. Valius and A. Kazlauskas, Phospholipase C-gamma 1 and phosphatidylinositol 3 kinase are the downstream mediators of the PDGF receptor's mitogenic signal, *Cell* **73** (1993) 321–334.
403. B.H. Jhun, D.W. Rose, B.L. Seely, L. Rameh, L. Cantley, A.R. Saltiel and J.M. Olefsky, Microinjection of the SH2 domain of the 85-kilodalton subunit of phosphatidylinositol 3-kinase inhibits insulin-induced DNA synthesis and c-fos expression, *Mol. Cell Biol.* **14** (1994) 7466–7475.
404. J. Glassford, E. Vigorito, I. Soeiro, P.A. Madureira, G. Zoumpoulidou, J.J. Brosens, M. Turner and E.W. Lam, Phosphatidylinositol 3-kinase is required for the transcriptional activation of cyclin D2 in BCR activated primary mouse B lymphocytes, *Eur. J. Immunol.* **35** (2005) 2748–2761.
405. S.J. Leevers, D. Weinkove, L.K. MacDougall, E. Hafen and M.D. Waterfield, The Drosophila phosphoinositide 3-kinase Dp110 promotes cell growth, *EMBO J.* **15** (1996) 6584–6594.

406. H. Zhang, J.P. Stallock, J.C. Ng, C. Reinhard and T.P. Neufeld, Regulation of cellular growth by the Drosophila target of rapamycin dTOR, *Genes Dev.* **14** (2000) 2712–2724.
407. C.H. Kwon, X. Zhu, J. Zhang, L.L. Knoop, R. Tharp, R.J. Smeyne, C.G. Eberhart, P.C. Burger and S.J. Baker, Pten regulates neuronal soma size: a mouse model of Lhermitte–Duclos disease, *Nat. Genet.* **29** (2001) 404–411.
408. C. Rommel, B.A. Clarke, S. Zimmermann, L. Nunez, R. Rossman, K. Reid, K. Moelling, G.D. Yancopoulos and D.J. Glass, Differentiation stage-specific inhibition of the Raf-MEK-ERK pathway by Akt, *Science* **286** (1999) 1738–1741.
409. M. Ihara-Watanabe, T. Uchihashi, Y. Miyauchi, N. Sakai, M. Yamagata, K. Ozono and T. Michigami, Involvement of phosphoinositide 3-kinase signaling pathway in chondrocytic differentiation of ATDC5 cells: application of a gene-trap mutagenesis, *J. Cell Biochem.* **93** (2004) 418–426.
410. T.R. Jackson, I.J. Blader, L.P. Hammonds-Odie, C.R. Burga, F. Cooke, P.T. Hawkins, A.G. Wolf, K.A. Heldman and A.B. Theibert, Initiation and maintenance of NGF-stimulated neurite outgrowth requires activation of a phosphoinositide 3-kinase, *J. Cell Sci.* **109** (Pt 2) (1996) 289–300.
411. J.S. Lee, Z. Hmama, A. Mui and N.E. Reiner, Stable gene silencing in human monocytic cell lines using lentiviral-delivered small interference RNA. Silencing of the p110alpha isoform of phosphoinositide 3-kinase reveals differential regulation of adherence induced by 1alpha,25-dihydroxycholecalciferol and bacterial lipopolysaccharide, *J. Biol. Chem.* **279** (2004) 9379–9388.
412. L.A. Greene and A.S. Tischler, Establishment of a noradrenergic clonal line of rat adrenal pheochromocytoma cells which respond to nerve growth factor, *Proc. Natl. Acad. Sci. USA* **73** (1976) 2424–2428.
413. P. Kaliman, F. Vinals, X. Testar, M. Palacin and A. Zorzano, Phosphatidylinositol 3-kinase inhibitors block differentiation of skeletal muscle cells, *J. Biol. Chem.* **271** (1996) 19146–19151.
414. A.M. Valverde, M. Lorenzo, P. Navarro and M. Benito, Phosphatidylinositol 3-kinase is a requirement for insulin-like growth factor I-induced differentiation, but not for mitogenesis, in fetal brown adipocytes, *Mol. Endocrinol.* **11** (1997) 595–607.
415. O. Halevy and L.C. Cantley, Differential regulation of the phosphoinositide 3-kinase and MAP kinase pathways by hepatocyte growth factors vs. insulin-like growth factor-I in myogenic cells, *Exp. Cell Res.* **297** (2004) 224–234.
416. E. Wandzioch, C.E. Edling, R.H. Palmer, L. Carlsson and B. Hallberg, Activation of the MAP kinase pathway by c-Kit is PI-3 kinase dependent in hematopoietic progenitor/stem cell lines, *Blood* **104** (2004) 51–57.

417. A. Ptasznik, G.M. Beattie, M.I. Mally, V. Cirulli, A. Lopez and A. Hayek, Phosphatidylinositol 3-kinase is a negative regulator of cellular differentiation, *J. Cell Biol.* **137** (1997) 1127–1136.
418. R. Baier, T. Bondeva, R. Klinger, A. Bondev and R. Wetzker, Retinoic acid induces selective expression of phosphoinositide 3-kinase gamma in myelomonocytic U937 cells, *Cell Growth Differ.* **10** (1999) 447–456.
419. V. Bertagnolo, L.M. Neri, M. Marchisio, C. Mischiati and S. Capitani, Phosphoinositide 3-kinase activity is essential for all-trans-retinoic acid-induced granulocytic differentiation of HL-60 cells, *Cancer Res.* **59** (1999) 542–546.
420. D. Visnjic, V. Crljen, J. Curic, D. Batinic, S. Volinia and H. Banfic, The activation of nuclear phosphoinositide 3-kinase C2beta in all-trans-retinoic acid-differentiated HL-60 cells, *FEBS Lett.* **529** (2002) 268–274.
421. A.J. Ridley, M.A. Schwartz, K. Burridge, R.A. Firtel, M.H. Ginsberg, G. Borisy, J.T. Parsons and A.R. Horwitz, Cell migration: integrating signals from front to back, *Science* **302** (2003) 1704–1709.
422. A.T. Sasaki, C. Chun, K. Takeda and R.A. Firtel, Localized Ras signaling at the leading edge regulates PI3K, cell polarity, and directional cell movement, *J. Cell Biol.* **167** (2004) 505–518.
423. S. Wennstrom, P. Hawkins, F. Cooke, K. Hara, K. Yonezawa, M. Kasuga, T. Jackson, L. Claesson-Welsh and L. Stephens, Activation of phosphoinositide 3-kinase is required for PDGF stimulated membrane ruffling, *Curr. Biol.* **4** (1994) 385–393.
424. J. Li, C. Yen, D. Liaw, K. Podsypanina, S. Bose, S.I. Wang, J. Puc, C. Miliaresis, L. Rodgers, R. McCombie, S.H. Bigner, B.C. Giovanella, M. Ittmann, B. Tycko, H. Hibshoosh, M.H. Wigler and R. Parsons, PTEN, a putative protein tyrosine phosphatase gene mutated in human brain, breast, and prostate cancer, *Science* **275** (1997) 1943–1947.
425. P.A. Steck, M.A. Pershouse, S.A. Jasser, W.K. Yung, H. Lin, A.H. Ligon, L.A. Langford, M.L. Baumgard, T. Hattier, T. Davis, C. Frye, R. Hu, B. Swedlund, D.H. Teng and S.V. Tavtigian, Identification of a candidate tumour suppressor gene, MMAC1, at chromosome 10q23.3 that is mutated in multiple advanced cancers, *Nat. Genet.* **15** (1997) 356–362.
426. D.M. Li and H. Sun, TEP1, encoded by a candidate tumor suppressor locus, is a novel protein tyrosine phosphatase regulated by transforming growth factor beta, *Cancer Res.* **57** (1997) 2124–2129.
427. D. Liaw, D.J. Marsh, J. Li, P.L. Dahia, S.I. Wang, Z. Zheng, S. Bose, K.M. Call, H.C. Tsou, M. Peacocke, C. Eng and R. Parsons, Germline mutations of the PTEN gene in Cowden disease, an inherited breast and thyroid cancer syndrome, *Nat. Genet.* **16** (1997) 64–67.

428. X.P. Zhou, D.J. Marsh, C.D. Morrison, A.R. Chaudhury, M. Maxwell, G. Reifenberger and C. Eng, Germline inactivation of PTEN and dysregulation of the phosphoinositol-3-kinase/Akt pathway cause human Lhermitte-Duclos disease in adults, *Am. J. Hum. Genet.* **73** (2003) 1191–1198.
429. M.R. Nelen, W.C. van Staveren, E.A. Peeters, M.B. Hassel, R.J. Gorlin, H. Hamm, C.F. Lindboe, J.P. Fryns, R.H. Sijmons, D.G. Woods, E.C. Mariman, G.W. Padberg and H. Kremer, Germline mutations in the PTEN/MMAC1 gene in patients with Cowden disease, *Hum. Mol. Genet.* **6** (1997) 1383–1387.
430. S.A. Backman, V. Stambolic, A. Suzuki, J. Haight, A. Elia, J. Pretorius, M.S. Tsao, P. Shannon, B. Bolon, G.O. Ivy and T.W. Mak, Deletion of Pten in mouse brain causes seizures, ataxia and defects in soma size resembling Lhermitte-Duclos disease, *Nat. Genet.* **29** (2001) 396–403.
431. F.B. Furnari, H. Lin, H.S. Huang and W.K. Cavenee, Growth suppression of glioma cells by PTEN requires a functional phosphatase catalytic domain, *Proc. Natl. Acad. Sci. USA* **94** (1997) 12479–12484.
432. C.A. Di, B. Pesce, C. Cordon-Cardo and P.P. Pandolfi, Pten is essential for embryonic development and tumour suppression, *Nat. Genet.* **19** (1998) 348–355.
433. M.E. McMenamin, P. Soung, S. Perera, I. Kaplan, M. Loda and W.R. Sellers, Loss of PTEN expression in paraffin-embedded primary prostate cancer correlates with high Gleason score and advanced stage, *Cancer Res.* **59** (1999) 4291–4296.
434. S.I. Wang, J. Puc, J. Li, J.N. Bruce, P. Cairns, D. Sidransky and R. Parsons, Somatic mutations of PTEN in glioblastoma multiforme, *Cancer Res.* **57** (1997) 4183–4186.
435. B.K. Rasheed, T.T. Stenzel, R.E. McLendon, R. Parsons, A.H. Friedman, H.S. Friedman, D.D. Bigner and S.H. Bigner, PTEN gene mutations are seen in high-grade but not in low-grade gliomas, *Cancer Res.* **57** (1997) 4187–4190.
436. J. Bostrom, J.M. Cobbers, M. Wolter, G. Tabatabai, R.G. Weber, P. Lichter, V.P. Collins and G. Reifenberger, Mutation of the PTEN (MMAC1) tumor suppressor gene in a subset of glioblastomas but not in meningiomas with loss of chromosome arm 10q, *Cancer Res.* **58** (1998) 29–33.
437. I.U. Ali, L.M. Schriml and M. Dean, Mutational spectra of PTEN/MMAC1 gene: a tumor suppressor with lipid phosphatase activity, *J. Natl. Cancer Inst.* **91** (1999) 1922–1932.
438. M. Iijima, Y.E. Huang, H.R. Luo, F. Vazquez and P.N. Devreotes, Novel mechanism of PTEN regulation by its phosphatidylinositol 4,5-bisphosphate

binding motif is critical for chemotaxis, *J. Biol. Chem.* **279** (2004) 16606–16613.

439. G. McConnachie, I. Pass, S.M. Walker and C.P. Downes, Interfacial kinetic analysis of the tumour suppressor phosphatase, PTEN: evidence for activation by anionic phospholipids, *Biochem. J.* **371** (2003) 947–955.
440. R.B. Campbell, F. Liu and A.H. Ross, Allosteric activation of PTEN phosphatase by phosphatidylinositol 4,5-bisphosphate, *J. Biol. Chem.* **278** (2003) 33617–33620.
441. J.O. Lee, H. Yang, M.M. Georgescu, C.A. Di, T. Maehama, Y. Shi, J.E. Dixon, P. Pandolfi and N.P. Pavletich, Crystal structure of the PTEN tumor suppressor: implications for its phosphoinositide phosphatase activity and membrane association, *Cell* **99** (1999) 323–334.
442. M. Raftopoulou, S. Etienne-Manneville, A. Self, S. Nicholls and A. Hall, Regulation of cell migration by the C2 domain of the tumor suppressor PTEN, *Science* **303** (2004) 1179–1181.
443. K.K. Dev, Making protein interactions druggable: targeting PDZ domains, *Nat. Rev. Drug Discov.* **3** (2004) 1047–1056.
444. N.B. Adey, L. Huang, P.A. Ormonde, M.L. Baumgard, R. Pero, D.V. Byreddy, S.V. Tavtigian and P.L. Bartel, Threonine phosphorylation of the MMAC1/PTEN PDZ binding domain both inhibits and stimulates PDZ binding, *Cancer Res.* **60** (2000) 35–37.
445. Y. Wu, D. Dowbenko, S. Spencer, R. Laura, J. Lee, Q. Gu and L.A. Lasky, Interaction of the tumor suppressor PTEN/MMAC with a PDZ domain of MAGI3, a novel membrane-associated guanylate kinase, *J. Biol. Chem.* **275** (2000) 21477–21485.
446. X. Wu, K. Hepner, S. Castelino-Prabhu, D. Do, M.B. Kaye, X.J. Yuan, J. Wood, C. Ross, C.L. Sawyers and Y.E. Whang, Evidence for regulation of the PTEN tumor suppressor by a membrane-localized multi-PDZ domain containing scaffold protein MAGI-2, *Proc. Natl. Acad. Sci. USA* **97** (2000) 4233–4238.
447. J. Torres and R. Pulido, The tumor suppressor PTEN is phosphorylated by the protein kinase CK2 at its C terminus. Implications for PTEN stability to proteasome-mediated degradation, *J. Biol. Chem.* **276** (2001) 993–998.
448. M.M. Georgescu, K.H. Kirsch, T. Akagi, T. Shishido and H. Hanafusa, The tumor-suppressor activity of PTEN is regulated by its carboxyl-terminal region, *Proc. Natl. Acad. Sci. USA* **96** (1999) 10182–10187.
449. F. Vazquez, S. Ramaswamy, N. Nakamura and W.R. Sellers, Phosphorylation of the PTEN tail regulates protein stability and function, *Mol. Cell Biol.* **20** (2000) 5010–5018.

450. F. Vazquez, S.R. Grossman, Y. Takahashi, M.V. Rokas, N. Nakamura and W.R. Sellers, Phosphorylation of the PTEN tail acts as an inhibitory switch by preventing its recruitment into a protein complex, *J. Biol. Chem.* **276** (2001) 48627–48630.
451. S. Das, J.E. Dixon and W. Cho, Membrane-binding and activation mechanism of PTEN, *Proc. Natl. Acad. Sci. USA* **100** (2003) 7491–7496.
452. C.P. Downes, S. Walker, G. McConnachie, Y. Lindsay, I.H. Batty and N.R. Leslie, Acute regulation of the tumour suppressor phosphatase, PTEN, by anionic lipids and reactive oxygen species, *Biochem. Soc. Trans.* **32** (2004) 338–342.
453. P. Zimmermann, K. Meerschaert, G. Reekmans, I. Leenaerts, J.V. Small, J. Vandekerckhove, G. David and J. Gettemans, PIP(2)-PDZ domain binding controls the association of syntenin with the plasma membrane, *Mol. Cell* **9** (2002) 1215–1225.
454. N.R. Leslie, A. Gray, I. Pass, E.A. Orchiston and C.P. Downes, Analysis of the cellular functions of PTEN using catalytic domain and C-terminal mutations: differential effects of C-terminal deletion on signalling pathways downstream of phosphoinositide 3-kinase, *Biochem. J.* **346** (Pt 3) (2000) 827–833.
455. N.R. Leslie, D. Bennett, A. Gray, I. Pass, K. Hoang-Xuan and C.P. Downes, Targeting mutants of PTEN reveal distinct subsets of tumour suppressor functions, *Biochem. J.* **357** (2001) 427–435.
456. I.W. Cheney, D.E. Johnson, M.T. Vaillan Court, J. Avanzini, A. Morimoto, G.W. Demers, K.N. Wills, P.W. Shabram, J.B. Bolen, S.V. Tavtigian, R. Bookstein, Suppression of tumorigenicity of glioblastoma cells by adenovirus-mediated MMAC1/PTEN gene transfer, *Cancer Res.* **58** (1998) 2331–2334.
457. D.M. Li and H. Sun, PTEN/MMAC1/TEP1 suppresses the tumorigenicity and induces G1 cell cycle arrest in human glioblastoma cells, *Proc. Natl. Acad. Sci. USA* **95** (1998) 15406–15411.
458. L. Li, B.R. Ernsting, M.J. Wishart, D.L. Lohse and J.E. Dixon, A family of putative tumor suppressors is structurally and functionally conserved in humans and yeast, *J. Biol. Chem.* **272** (1997) 29403–29406.
459. M.P. Myers, J.P. Stolarov, C. Eng, J. Li, S.I. Wang, M.H. Wigler, R. Parsons and N.K. Tonks, P-TEN, the tumor suppressor from human chromosome 10q23, is a dual-specificity phosphatase, *Proc. Natl. Acad. Sci. USA* **94** (1997) 9052–9057.
460. M. Tamura, J. Gu, K. Matsumoto, S. Aota, R. Parsons and K.M. Yamada, Inhibition of cell migration, spreading, and focal adhesions by tumor suppressor PTEN, *Science* **280** (1998) 1614–1617.

461. M. Tamura, J. Gu, E.H. Danen, T. Takino, S. Miyamoto and K.M. Yamada, PTEN interactions with focal adhesion kinase and suppression of the extracellular matrix-dependent phosphatidylinositol 3-kinase/Akt cell survival pathway, *J. Biol. Chem.* **274** (1999) 20693–20703.
462. J. Gu, M. Tamura, R. Pankov, E.H. Danen, T. Takino, K. Matsumoto and K.M. Yamada, Shc and FAK differentially regulate cell motility and directionality modulated by PTEN, *J. Cell Biol.* **146** (1999) 389–403.
463. M.P. Myers, I. Pass, I.H. Batty, K.J. Van der, J.P. Stolarov, B.A. Hemmings, M.H. Wigler, C.P. Downes and N.K. Tonks, The lipid phosphatase activity of PTEN is critical for its tumor supressor function, *Proc. Natl. Acad. Sci. USA* **95** (1998) 13513–13518.
464. M.P. Myers and N.K. Tonks, PTEN: Sometimes taking it off can be better than putting it on, *Am. J. Hum. Genet.* **61** (1997) 1234–1238.
465. T. Maehama and J. Dixon, The tumor suppressor, PTEN/MMAC1, dephosphorylates the lipid second messenger, phosphatidylinositol 3,4,5-trisphosphate, *J. Biol. Chem.* **273** (1998) 13375–13378.
466. V. Stambolic, A. Suzuki, J. de la Pompa, G. Brothers, C. Mirtsos, T. Sasaki, J. Ruland, J. Penninger, D. Siderovski and T. Mak, Negative regulation of PKB/Akt-dependent cell survival by the tumor suppressor PTEN, *Cell* **95** (1998) 29–39.
467. D. Haas-Kogan, N. Shalev, M. Wong, G. Mills, G. Yount and D. Stokoe, Protein kinase B (PKB/Akt) activity is elevated in glioblastoma cells due to mutation of the tumor suppressor PTEN/MMAC, *Curr. Biol.* **8** (1998) 1195–1198.
468. F. Solari, A. Bourbon-Piffaut, I. Masse, B. Payrastre, A.M. Chan and M. Billaud, The human tumour suppressor PTEN regulates longevity and dauer formation in Caenorhabditis elegans, *Oncogene* **24** (2005) 20–27.
469. S. Ramaswamy, N. Nakamura, F. Vazquez, D.B. Batt, S. Perera, T.M. Roberts and W.R. Sellers, Regulation of G1 progression by the PTEN tumor suppressor protein is linked to inhibition of the phosphatidylinositol 3-kinase/Akt pathway, *Proc. Natl. Acad. Sci. USA* **96** (1999) 2110–2115.
470. J.M. Paramio, M. Navarro, C. Segrelles, E. Gomez-Casero and J.L. Jorcano, PTEN tumour suppressor is linked to the cell cycle control through the retinoblastoma protein, *Oncogene* **18** (1999) 7462–7468.
471. J. Torres, J. Rodriguez, M.P. Myers, M. Valiente, J.D. Graves, N.K. Tonks and R. Pulido, Phosphorylation regulated cleavage of the tumor suppressor PTEN by caspase-3: implications for the control of protein stability and PTEN-protein interactions, *J. Biol. Chem.* **278** (2003) 30652–30660.

472. F. Okahara, H. Ikawa, Y. Kanaho and T. Machama, Regulation of PTEN phosphorylation and stability by a tumour suppressor candidate protein, *J. Biol. Chem.* **279** (2004) 45300–45303.
473. M. Valiente, A. Andres-Pons, B. Gomar, J. Torres, A. Gil, C. Tapparel, S.E. Antonarakis and R. Pulido, Binding of PTEN to specific PD2 domains contributes to PTEN stability and phosphorylation by microtubule associated serine/threonine kinases, *J. Biol. Chem.* **280** (2005) 28936–28943.
474. A.E. Lehesjoki, E.M. Sankila, J. Miao, M. Somer, R. Salonen, J. Rapola and C.A. de la, X linked neonatal myotubular myopathy: one recombination detected with four polymorphic DNA markers from Xq28, *J. Med. Genet.* **27** (1990) 288–291.
475. C. Wallgren-Pettersson, A. Clarke, F. Samson, M. Fardeau, V. Dubowitz, H. Moser, T. Grimm, R.J. Barohn and P.G. Barth, The myotubular myopathies: differential diagnosis of the X linked recessive, autosomal dominant, and autosomal recessive forms and present state of DNA studies, *J. Med. Genet.* **32** (1995) 673–679.
476. J.A. Sawchak, J.H. Sher, M.G. Norman, R.W. Kula and S.A. Shafiq, Centronuclear myopathy heterogeneity: distinction of clinical types by myosin isoform patterns, *Neurology* **41** (1991) 135–140.
477. A. Fidzianska, I. Warlo and H.H. Goebel, Neonatal centronuclear myopathy with N-CAM decorated myotubes, *Neuropediatrics* **25** (1994) 158–161.
478. J. Schaletzky, S.K. Dove, B. Short, O. Lorenzo, M.J. Clague and F.A. Barr, Phosphatidylinositol-5-phosphate activation and conserved substrate specificity of the myotubularin phosphatidylinositol 3-phosphatases, *Curr. Biol.* **13** (2003) 504–509.
479. J. Laporte, F. Bedez, A. Bolino and J.L. Mandel, Myotubularins, a large disease-associated family of cooperating catalytically active and inactive phosphoinositides phosphatases, *Hum. Mol. Genet.* **12** (Spec no. 2) (2003) R285–R292.
480. A. Bolino, M. Muglia, F.L. Conforti, E. LeGuern, M.A. Salih, D.M. Georgiou, K. Christodoulou, I. Hausmanowa-Petrusewicz, P. Mandich, A. Schenone, A. Gambardella, F. Bono, A. Quattrone, M. Devoto and A.P. Monaco, Charcot–Marie–Tooth type 4B is caused by mutations in the gene encoding myotubularin-related protein-2, *Nat. Genet.* **25** (2000) 17–19.
481. A. Bolino, V. Marigo, F. Ferrera, J. Loader, L. Romio, A. Leoni, D.M. Di, R. Cinti, C. Cecchi, M.L. Feltri, L. Wrabetz, R. Ravazzolo and A.P. Monaco, Molecular characterization and expression analysis of Mtmr2, mouse homologue of MTMR2, the Myotubularin-related 2 gene, mutated in CMT4B, *Gene* **283** (2002) 17–26.

482. R. Firestein, P.L. Nagy, M. Daly, P. Huie, M. Conti and M.L. Cleary, Male infertility, impaired spermatogenesis, and azoospermia in mice deficient for the pseudophosphatase Sbf1, *J. Clin. Invest.* **109** (2002) 1165–1172.
483. X. Cui, V. De, I, R. Slany, A. Miyamoto, R. Firestein and M.L. Cleary, Association of SET domain and myotubularin-related proteins modulates growth control, *Nat. Genet.* **18** (1998) 331–337.
484. M.J. Begley, G.S. Taylor, S.A. Kim, D.M. Veine, J.E. Dixon and J.A. Stuckey, Crystal structure of a phosphoinositide phosphatase, MTMR2: insights into myotubular myopathy and Charcot-Marie-Tooth syndrome, *Mol. Cell* **12** (2003) 1391–1402.
485. T. Doerks, M. Strauss, M. Brendel and P. Bork, GRAM, a novel domain in glucosyltransferases, myotubularins and other putative membrane-associated proteins, *Trends Biochem. Sci.* **25** (2000) 483–485.
486. Y. Mochizuki and P.W. Majerus, Characterization of myotubularin-related protein 7 and its binding partner, myotubularin-related protein 9, *Proc. Natl. Acad. Sci. USA* **100** (2003) 9768–9773.
487. S.A. Kim, P.O. Vacratsis, R. Firestein, M.L. Cleary and J.E. Dixon, Regulation of myotubularin-related (MTMR)2 phosphatidylinositol phosphatase by MTMR5, a catalytically inactive phosphatase, *Proc. Natl. Acad. Sci. USA* **100** (2003) 4492–4497.
488. O. Lorenzo, S. Urbe and M.J. Clague, Analysis of phosphoinositide binding domain properties within the myotubularin-related protein MTMR3, *J. Cell Sci.* **118** (2005) 2005–2012.
489. P. Berger, C. Schaffitzel, I. Berger, N. Ban and U. Suter, Membrane association of myotubularin-related protein 2 is mediated by a pleckstrin homology-GRAM domain and a coiled-coil dimerization module, *Proc. Natl. Acad. Sci. USA* **100** (2003) 12177–12182.
490. G.S. Taylor, T. Maehama and J.E. Dixon, Inaugural article: myotubularin, a protein tyrosine phosphatase mutated in myotubular myopathy, dephosphorylates the lipid second messenger, phosphatidylinositol 3-phosphate, *Proc. Natl. Acad. Sci. USA* **97** (2000) 8910–8915.
491. L. Liu, J.E. Damen, R.L. Cutler and G. Krystal, Multiple cytokines stimulate the binding of a common 145-kilodalton protein to Shc at the Grb2 recognition site of Shc, *Mol. Cell Biol.* **14** (1994) 6926–6935.
492. J.E. Damen, L. Liu, P. Rosten, R.K. Humphries, A.B. Jefferson, P.W. Majerus and G. Krystal, The 145-kDa protein induced to associate with Shc by multiple cytokines is an inositol tetraphosphate and phosphatidylinositol 3,4,5-triphosphate 5-phosphatase, *Proc. Natl. Acad. Sci. USA* **93** (1996) 1689–1693.

493. Z. Tu, J.M. Ninos, Z. Ma, J.W. Wang, M.P. Lemos, C. Desponts, T. Ghansah, J.M. Howson and W.G. Kerr, Embryonic and hematopoietic stem cells express a novel SH2-containing inositol 5′-phosphatase isoform that partners with the Grb2 adapter protein, *Blood* **98** (2001) 2028–2038.
494. X. Pesesse, C. Moreau, A.L. Drayer, R. Woscholski, P. Parker and C. Erneux, The SH2 domain containing inositol 5-phosphatase SHIP2 displays phosphatidylinositol 3,4,5-trisphosphate and inositol 1,3,4,5-tetrakisphosphate 5-phosphatase activity, *FEBS Lett.* **437** (1998) 301–303.
495. J.E. Damen, L. Liu, M.D. Ware, M. Ermolaeva, P.W. Majerus and G. Krystal, Multiple forms of the SH2-containing inositol phosphatase, SHIP, are generated by C-terminal truncation, *Blood* **92** (1998) 1199–1205.
496. S. Horn, J. Meyer, J. Heukeshoven, B. Fehse, C. Schulze, S. Li, J. Frey, S. Poll, C. Stocking and M. Jucker, The inositol 5-phosphatase SHIP is expressed as 145 and 135 kDa proteins in blood and bone marrow cells in vivo, whereas carboxyl-truncated forms of SHIP are generated by proteolytic cleavage in vitro, *Leukemia* **15** (2001) 112–120.
497. M.A. Osborne, G. Zenner, M. Lubinus, X. Zhang, Z. Songyang, L.C. Cantley, P. Majerus, P. Burn and J.P. Kochan, The inositol 5′-phosphatase SHIP binds to immunoreceptor signaling motifs and responds to high affinity IgE receptor aggregation, *J. Biol. Chem.* **271** (1996) 29271–29278.
498. M. Huber, C.D. Helgason, J.E. Damen, M. Scheid, V. Duronio, L. Liu, M.D. Ware, R.K. Humphries and G. Krystal, The role of SHIP in growth factor induced signalling, *Prog. Biophys. Mol. Biol.* **71** (1999) 423–434.
499. M. Sattler, S. Verma, Y.B. Pride, R. Salgia, L.R. Rohrschneider and J.D. Griffin, SHIP1, an SH2 domain containing polyinositol-5-phosphatase, regulates migration through two critical tyrosine residues and forms a novel signaling complex with DOK1 and CRKL, *J. Biol. Chem.* **276** (2001) 2451–2458.
500. D. Wisniewski, A. Strife, S. Swendeman, H. Erdjument-Bromage, S. Geromanos, W.M. Kavanaugh, P. Tempst and B. Clarkson, A novel SH2-containing phosphatidylinositol 3,4,5-trisphosphate 5-phosphatase (SHIP2) is constitutively tyrosine phosphorylated and associated with src homologous and collagen gene (SHC) in chronic myelogenous leukemia progenitor cells, *Blood* **93** (1999) 2707–2720.
501. H. Phee, A. Jacob and K.M. Coggeshall, Enzymatic activity of the Src homology 2 domain-containing inositol phosphatase is regulated by a plasma membrane location, *J. Biol. Chem.* **275** (2000) 19090–19097.
502. L. Velazquez, G.D. Gish, G.P. van Der, L. Taylor, J. Shulman and T. Pawson, The shc adaptor protein forms interdependent phosphotyrosine-mediated

protein complexes in mast cells stimulated with interleukin 3, *Blood* **96** (2000) 132–138.

503. L. Liu, J.E. Damen, M.R. Hughes, I. Babic, F.R. Jirik and G. Krystal, The Src homology 2 (SH2) domain of SH2-containing inositol phosphatase (SHIP) is essential for tyrosine phosphorylation of SHIP, its association with Shc, and its induction of apoptosis, *J. Biol. Chem.* **272** (1997) 8983–8988.
504. T.D. Lamkin, S.F. Walk, L. Liu, J.E. Damen, G. Krystal and K.S. Ravichandran, Shc interaction with Src homology 2 domain containing inositol phosphatase (SHIP) in vivo requires the Shc-phosphotyrosine binding domain and two specific phosphotyrosines on SHIP, *J. Biol. Chem.* **272** (1997) 10396–10401.
505. Y. Liu, B. Jenkins, J.L. Shin and L.R. Rohrschneider, Scaffolding protein Gab2 mediates differentiation signaling downstream of Fms receptor tyrosine kinase, *Mol. Cell. Biol.* **21** (2001) 3047–3056.
506. S. Lemay, D. Davidson, S. Latour and A. Veillette, Dok-3, a novel adapter molecule involved in the negative regulation of immunoreceptor signaling, *Mol. Cell. Biol.* **20** (2000) 2743–2754.
507. J.E. Damen, M.D. Ware, J. Kalesnikoff, M.R. Hughes and G. Krystal, SHIP's C-terminus is essential for its hydrolysis of PIP3 and inhibition of mast cell degranulation, *Blood* **97** (2001) 1343–1351.
508. S. Marchetto, E. Fournier, N. Beslu, T. urran-Schleinitz, P. Dubreuil, J.P. Borg, D. Birnbaum and O. Rosnet, SHC and SHIP phosphorylation and interaction in response to activation of the FLT3 receptor, *Leukemia* **13** (1999) 1374–1382.
509. S. Tridandapani, T. Kelley, M. Pradhan, D. Cooney, L.B. Justement and K.M. Coggeshall, Recruitment and phosphorylation of SH2-containing inositol phosphatase and Shc to the B-cell Fc gamma immunoreceptor tyrosine-based inhibition motif peptide motif, *Mol. Cell. Biol.* **17** (1997) 4305–4311.
510. T. Kimura, H. Sakamoto, E. Appella and R.P. Siraganian, The negative signaling molecule SH2 domain-containing inositol-polyphosphate 5-phosphatase (SHIP) binds to the tyrosine-phosphorylated beta subunit of the high affinity IgE receptor, *J. Biol. Chem.* **272** (1997) 13991–13996.
511. S. Tridandapani, Y. Wang, C.B. Marsh and C.L. Anderson, Src homology 2 domain-containing inositol polyphosphate phosphatase regulates NF-kappa B-mediated gene transcription by phagocytic Fc gamma Rs in human myeloid cells, *J. Immunol.* **169** (2002) 4370–4378.
512. C.D. Helgason, J.E. Damen, P. Rosten, R. Grewal, P. Sorensen, S.M. Chappel, A. Borowski, F. Jirik, G. Krystal and R.K. Humphries, Targeted disruption

of SHIP leads to hemopoietic perturbations, lung pathology, and a shortened life span, *Genes Dev.* **12** (1998) 1610–1620.

513. S. Takeshita, N. Namba, J.J. Zhao, Y. Jiang, H.K. Genant, M.J. Silva, M.D. Brodt, C.D. Helgason, J. Kalesnikoff, M.J. Rauh, R.K. Humphries, G. Krystal, S.L. Teitelbaum and F.P. Ross, SHIP-deficient mice are severely osteoporotic due to increased numbers of hyper-resorptive osteoclasts, *Nat. Med.* **8** (2002) 943–949.
514. J.L. Moody, L. Xu, C.D. Helgason and F.R. Jirik, Anemia, thrombocytopenia, leukocytosis, extramedullary hematopoiesis, and impaired progenitor function in Pten+/−SHIP−/− mice: a novel model of myelodysplasia, *Blood* **103** (2004) 4503–4510.
515. S. Clement, U. Krause, F. Desmedt, J.F. Tanti, J. Behrends, X. Pesesse, T. Sasaki, J. Penninger, M. Doherty, W. Malaisse, J.E. Dumont, Y. Le Marchand-Brustel, C. Erneux, L. Hue and S. Schurmans, The lipid phosphatase SHIP2 controls insulin sensitivity, *Nature* **409** (2001) 92–97.
516. E. Marion, P.J. Kaisaki, V. Pouillon, C. Gueydan, J.C. Levy, A. Bodson, G. Krzentowski, J.C. Daubresse, J. Mockel, J. Behrends, G. Servais, C. Szpirer, V. Kruys, D. Gauguier and S. Schurmans, The gene INPPL1, encoding the lipid phosphatase SHIP2, is a candidate for type 2 diabetes in rat and man, *Diabetes* **51** (2002) 2012–2017.
517. H.W. Chang, M. Aoki, D. Fruman, K.R. Auger, A. Bellacosa, P.N. Tsichlis, L.C. Cantley, T.M. Roberts and P.K. Vogt, Transformation of chicken cells by the gene encoding the catalytic subunit of PI 3-kinase, *Science* **276** (1997) 1848–1850.
518. S. Kang, A. Denley, B. Vanhaesebroeck and P.K. Vogt, Oncogenic transformation induced by the pllobeta, -gamma and -delta isoforms of class I phosphoinositide 3-kinase, *Proc. Natl. Acad. Sci.* **103** (2006) 1289–1294.
519. Y.Y. Ma, S.J. Wei, Y.C. Lin, J.C. Lung, T.C. Chang, J. Whang-Peng, J.M. Liu, D.M. Yang, W.K. Yang and C.Y. Shen, PIK3CA as an oncogene in cervical cancer, *Oncogene* **19** (2000) 2739–2744.
520. L. Shayesteh, Y. Lu, W.L. Kuo, R. Baldocchi, T. Godfrey, C. Collins, D. Pinkel, B. Powell, G.B. Mills and J.W. Gray, PIK3CA is implicated as an oncogene in ovarian cancer, *Nat. Genet.* **21** (1999) 99–102.
521. G. Wu, E. Mambo, Z. Guo, S. Hu, X. Huang, S.M. Gollin, B. Trink, P.W. Ladenson, D. Sidransky and M. Xing, Uncommon mutation but common amplifications of the PIK3CA gene in thyroid tumors, *J. Clin. Endocrinol. Metab.* **90** (2005) 4688–4693.
522. Y. Samuels, Z. Wang, A. Bardelli, N. Silliman, J. Ptak, S. Szabo, H. Yan, A. Gazdar, S.M. Powell, G.J. Riggins, J.K. Willson, S. Markowitz, K.W.

Kinzler, B. Vogelstein and V.E. Velculescu, High frequency of mutations of the PIK3CA gene in human cancers, *Science* **304** (2004) 554.
523. R. Redon, D. Muller, K. Caulee, K. Wanherdrick, J. Abecassis and M.S. du, A simple specific pattern of chromosomal aberrations at early stages of head and neck squamous cell carcinomas: PIK3CA but not p63 gene as a likely target of 3q26-qter gains, *Cancer Res.* **61** (2001) 4122–4129.
524. K.E. Bachman, P. Argani, Y. Samuels, N. Silliman, J. Ptak, S. Szabo, H. Konishi, B. Karakas, B.G. Blair, C. Lin, B.A. Peters, V.E. Velculescu and B.H. Park, The PIK3CA gene is mutated with high frequency in human breast cancers, *Cancer Biol. Ther.* **3** (2004) 772–775.
525. A. Bardelli and V.E. Velculescu, Mutational analysis of gene families in human cancer, *Curr. Opin. Genet. Dev.* **15** (2005) 5–12.
526. C. Jimenez, D.R. Jones, P. Rodriguez-Viciana, A. Gonzalez-Garcia, E. Leonardo, S. Wennstrom, K.C. von, J.L. Toran, L. Borlado, V. Calvo, S.G. Copin, J.P. Albar, M.L. Gaspar, E. Diez, M.A. Marcos, J. Downward, A. Martinez, I. Merida and A.C. Carrera, Identification and characterization of a new oncogene derived from the regulatory subunit of phosphoinositide 3-kinase, *EMBO J.* **17** (1998) 743–753.
527. L.R. Borlado, C. Redondo, B. Alvarez, C. Jimenez, L.M. Criado, J. Flores, M.A. Marcos, A. Martinez, D. Balomenos and A.C. Carrera, Increased phosphoinositide 3-kinase activity induces a lymphoproliferative disorder and contributes to tumor generation in vivo, *FASEB J.* **14** (2000) 895–903.
528. W.A. Phillips, C.F. St, A.D. Munday, R.J. Thomas and C.A. Mitchell, Increased levels of phosphatidylinositol 3-kinase activity in colorectal tumors, *Cancer* **83** (1998) 41–47.
529. A.J. Philp, I.G. Campbell, C. Leet, E. Vincan, S.P. Rockman, R.H. Whitehead, R.J. Thomas and W.A. Phillips, The phosphatidylinositol 3′-kinase p85alpha gene is an oncogene in human ovarian and colon tumors, *Cancer Res.* **61** (2001) 7426–7429.
530. K. Nakanishi, M. Sakamoto, J. Yasuda, M. Takamura, N. Fujita, T. Tsuruo, S. Todo and S. Hirohashi, Critical involvement of the phosphatidylinositol 3-kinase/Akt pathway in anchorage-independent growth and hematogeneous intrahepatic metastasis of liver cancer, *Cancer Res.* **62** (2002) 2971–2975.
531. L.M. Shaw, I. Rabinovitz, H.H. Wang, A. Toker and A.M. Mercurio, Activation of phosphoinositide 3-OH kinase by the alpha6beta4 integrin promotes carcinoma invasion, *Cell* **91** (1997) 949–960.
532. Z. Yue, S. Jin, C. Yang, A.J. Levine and N. Heintz, Beclin 1, an autophagy gene essential for early embryonic development, is a haploinsufficient tumor suppressor, *Proc. Natl. Acad. Sci. USA* **100** (2003) 15077–15082.

533. M.S. Neshat, I.K. Mellinghoff, C. Tran, B. Stiles, G. Thomas, R. Petersen, P. Frost, J.J. Gibbons, H. Wu and C.L. Sawyers, Enhanced sensitivity of PTEN-deficient tumors to inhibition of FRAP/mTOR, *Proc. Natl. Acad. Sci. USA* **98** (2001) 10314–10319.
534. K. Podsypanina, R.T. Lee, C. Politis, I. Hennessy, A. Crane, J. Puc, M. Neshat, H. Wang, L. Yang, J. Gibbons, P. Frost, V. Dreisbach, J. Blenis, Z. Gaciong, P. Fisher, C. Sawyers, L. Hedrick-Ellenson and R. Parsons, An inhibitor of mTOR reduces neoplasia and normalizes p70/S6 kinase activity in Pten+/− mice, *Proc. Natl. Acad. Sci. USA* **98** (2001) 10320–10325.
535. M. Aoki, O. Batista, A. Bellacosa, P. Tsichlis and P.K. Vogt, The akt kinase: molecular determinants of oncogenicity, *Proc. Natl. Acad. Sci. USA* **95** (1998) 14950–14955.
536. C.B. Knobbe and G. Reifenberger, Genetic alterations and aberrant expression of genes related to the phosphatidyl-inositol-3′-kinase/protein kinase B (Akt) signal transduction pathway in glioblastomas, *Brain Pathol.* **13** (2003) 507–518.
537. S.P. Staal, Molecular cloning of the akt oncogene and its human homologues AKT1 and AKT2: amplification of AKT1 in a primary human gastric adenocarcinoma, *Proc. Natl. Acad. Sci. USA* **84** (1987) 5034–5037.
538. A. Bellacosa, F.D. de, A.K. Godwin, D.W. Bell, J.Q. Cheng, D.A. Altomare, M. Wan, L. Dubeau, G. Scambia, V. Masciullo, *et al.*, Molecular alterations of the AKT2 oncogene in ovarian and breast carcinomas, *Int. J. Cancer* **64** (1995) 280–285.
539. J.Q. Cheng, B. Ruggeri, W.M. Klein, G. Sonoda, D.A. Altomare, D.K. Watson and J.R. Testa, Amplification of AKT2 in human pancreatic cells and inhibition of AKT2 expression and tumorigenicity by antisense RNA, *Proc. Natl. Acad. Sci. USA* **93** (1996) 3636–3641.
540. Z.Q. Yuan, M. Sun, R.I. Feldman, G. Wang, X. Ma, C. Jiang, D. Coppola, S.V. Nicosia and J.Q. Cheng, Frequent activation of AKT2 and induction of apoptosis by inhibition of phosphoinositide-3-OH kinase/Akt pathway in human ovarian cancer, *Oncogene* **19** (2000) 2324–2330.
541. K. Nakatani, D.A. Thompson, A. Barthel, H. Sakaue, W. Liu, R.J. Weigel and R.A. Roth, Up-regulation of Akt3 in estrogen receptor-deficient breast cancers and androgen-independent prostate cancer lines, *J. Biol. Chem.* **274** (1999) 21528–21532.
542. I. Vivanco and C.L. Sawyers, The phosphatidylinositol 3-Kinase AKT pathway in human cancer, *Nat. Rev. Cancer* **2** (2002) 489–501.
543. S. Tanno, N. Yanagawa, A. Habiro, K. Koizumi, Y. Nakano, M. Osanai, Y. Mizukami, T. Okumura, J.R. Testa and Y. Kohgo, Serine/threonine kinase

AKT is frequently activated in human bile duct cancer and is associated with increased radioresistance, *Cancer Res.* **64** (2004) 3486–3490.

544. S. Yamamoto, Y. Tomita, Y. Hoshida, T. Morooka, H. Nagano, K. Dono, K. Umeshita, M. Sakon, O. Ishikawa, H. Ohigashi, S. Nakamori, M. Monden and K. Aozasa, Prognostic significance of activated Akt expression in pancreatic ductal adenocarcinoma, *Clin. Cancer Res.* **10** (2004) 2846–2850.
545. G. Perez-Tenorio and O. Stal, Activation of AKT/PKB in breast cancer predicts a worse outcome among endocrine treated patients, *Br. J. Cancer* **86** (2002) 540–545.
546. R.P. Ermoian, C.S. Furniss, K.R. Lamborn, D. Basila, M.S. Berger, A.R. Gottschalk, M.K. Nicholas, D. Stokoe and D.A. Haas-Kogan, Dysregulation of PTEN and protein kinase B is associated with glioma histology and patient survival, *Clin. Cancer Res.* **8** (2002) 1100–1106.
547. K. Nakanishi, M. Sakamoto, S. Yamasaki, S. Todo and S. Hirohashi, Akt phosphorylation is a risk factor for early disease recurrence and poor prognosis in hepatocellular carcinoma, *Cancer* **103** (2005) 307–312.
548. F.B. Furnari, H.J. Huang and W.K. Cavenee, The phosphoinositol phosphatase activity of PTEN mediates a serum-sensitive G1 growth arrest in glioma cells, *Cancer Res.* **58** (1998) 5002–5008.
549. W. Wick, F.B. Furnari, U. Naumann, W.K. Cavenee and M. Weller, PTEN gene transfer in human malignant glioma: sensitization to irradiation and CD95L induced apoptosis, *Oncogene* **18** (1999) 3936–3943.
550. L.P. Weng, W.M. Smith, P.L. Dahia, U. Ziebold, E. Gil, J.A. Lees and C. Eng, PTEN suppresses breast cancer cell growth by phosphatase activity-dependent G1 arrest followed by cell death, *Cancer Res.* **59** (1999) 5808–5814.
551. M.A. Davies, D. Koul, H. Dhesi, R. Berman, T.J. McDonnell, D. McConkey, W.K. Yung and P.A. Steck, Regulation of Akt/PKB activity, cellular growth, and apoptosis in prostate carcinoma cells by MMAC/PTEN, *Cancer Res.* **59** (1999) 2551–2556.
552. Y. Lu, Y.Z. Lin, R. LaPushin, B. Cuevas, X. Fang, S.X. Yu, M.A. Davies, H. Khan, T. Furui, M. Mao, R. Zinner, M.C. Hung, P. Steck, K. Siminovitch and G.B. Mills, The PTEN/MMAC1/TEP tumor suppressor gene decreases cell growth and induces apoptosis and anoikis in breast cancer cells, *Oncogene* **18** (1999) 7034–7045.
553. L. Rodriguez-Borlado, D.F. Barber, C. Hernandez, M.A. Rodriguez-Marcos, A. Sanchez, E. Hirsch, M. Wymann, A. Martinez and A.C. Carrera, Phosphatidylinositol 3-kinase regulates the CD4/CD8 T cell differentiation ratio, *J. Immunol.* **170** (2003) 4475–4482.

554. K. Okkenhaug and B. Vanhaesebroeck, PI3K signalling in B- and T-cells insights from gene-targeted mice, *Biochem. Soc. Trans.* **31** (2003) 270–274.
555. A. Bilancio, K. Okkenhaug, M. Camps, J.L. Emery, T. Ruckle, C. Rommel and B. Vanhaesebroeck, Key role of the p110 delta isoform of PI3K in B-cell antigen and IL4 receptor: comparative analysis of genetic and pharmacologic interference with p110 delta function in B-cells, *Blood* **107** (2006) 642–650.
556. F. Vigorito, G. Bardi, J. Glassford, E.W. Lam, E. Clayton and M. Turner, Vav-dependent and Vav-independent phosphatidylinositol 3-kinase activation in murine B-cells determined by the nature of the stimulus, *J. Immunol.* **173** (2004) 3209–3314.
557. W. Swat, V. Montgrain, T.A. Doggett, J. Donangpanya, K. Puri, W. Vermi, and T.G. Dracovo, The essential role of PI3K (delta) and PI3K (gamma) in thymocyte survival, *Blood* 2005 November 22 (Epub ahead of print).
558. A. Di Cristofano, P. Kotsi, Y. Peng, C. Cordon-Cardo, K. Elkon and P. Pandolfi, Impaired Fas response and autoimmunity in Pten+/− mice, *Science* **285** (1999) 2122–2125.

4 PI3K signalling to cell survival and proliferation

Paul Brennan

1. Molecules Directly Regulated By D3 Lipids

Phosphoinositol 3-kinase (PI3K) regulates a wide range of cellular processes through a complicated network of molecules (Fig. 1). In gaining a deeper understanding of these processes, it is wise to begin at the products of the enzyme PI3K, which are lipids phosphorylated at the D3 position of the inositol headgroup of inositol phospholipids. Any functional or molecular network regulated by PI3K activity must, somehow, be affected by these lipids. The discovery of high affinity binding of the protein PH domain to D3 phospholipids was a key step in our understanding of the molecular events following PI3K activity. Most of the molecules regulated by PI3K are the result of altered activation of a small number of PH domain-containing proteins.

The PH domain is a globular domain of about 100 residues (Interpro number: IPR001849). This domain is found in more than 500 proteins in the human genome and is shared throughout the animal and plant kingdom. However, not all PH domains demonstrate specificity for D3 lipids. The residues involved in high affinity binding to D3 phosphorylated phosphoinositides lie at the N-terminus of the domain and contains a $KX_{(7-13)}R/KXR^*$ (where X is any amino acid and * is a hydrophobic amino acid).[1–3] Substantially fewer proteins contain this motif and here four of the most important PH-containing protein families that have been

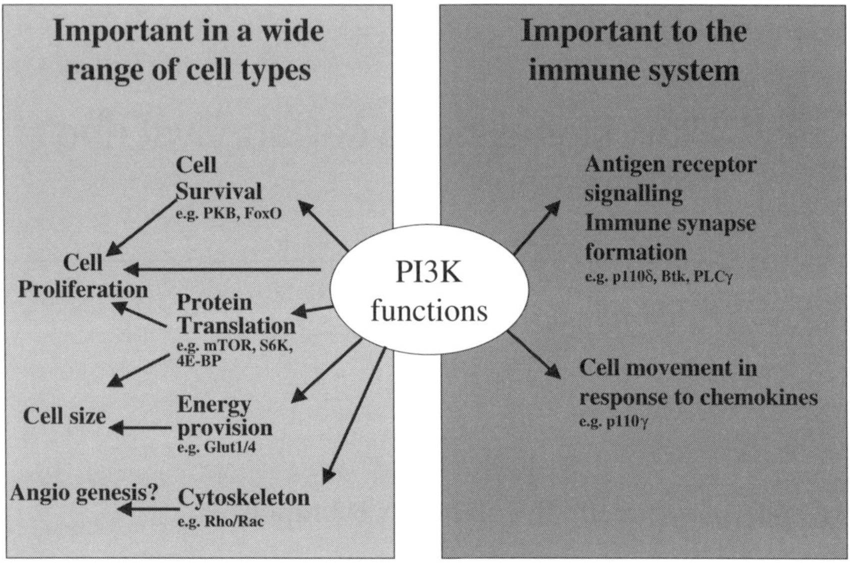

Fig. 1 A simplified schematic of some important PI3K functions.

comprehensively studied and demonstrated to be regulated by PI3K are described. The four families are the PKB/Akt family, the molecule PDK1, the TEC kinase family and the phospholipase C family.

In many cases, D3 lipids appear to act as a type of molecular scaffold to bring together kinases and their substrates. This is the case for PDK1 and PKB/Akt. Often, the D3 lipid molecular scaffold seems to cooperate with protein scaffolds or adaptor molecules to bring together a multiprotein complex. For example, the recruitment of Btk and PLCγ1 following antigen receptor signalling requires both D3 lipids and a platform of tyrosine phosphorylated proteins. This complex is called the antigen receptor signalosome and requires PI3K activity for proper assembly (Fig. 2).

In this chapter, some of the key molecules regulated directly by D3 lipids and some molecules downstream of these are described. There will be information derived from various model organisms including the worm, *C. elegans*, the fruit fly, *Drosophila melanogaster*, and the mouse, *Mus musculus*.[3,4] There is now clear evidence that the pathway leading from tyrosine kinase receptors, through PI3K leading to the activation of the serine kinases PDK-1 and PKB/Akt is evolutionarily conserved. One conserved

Name	Entry Name	Accession No.	No. of amino acids
PDK1	PDPK1_HUMAN	O15530	556
Btk	BTK_HUMAN	Q06187	659
PKB/Akt	AKT1_HUMAN	P31749	480
S6K	KS6B1_HUMAN	P23443	525

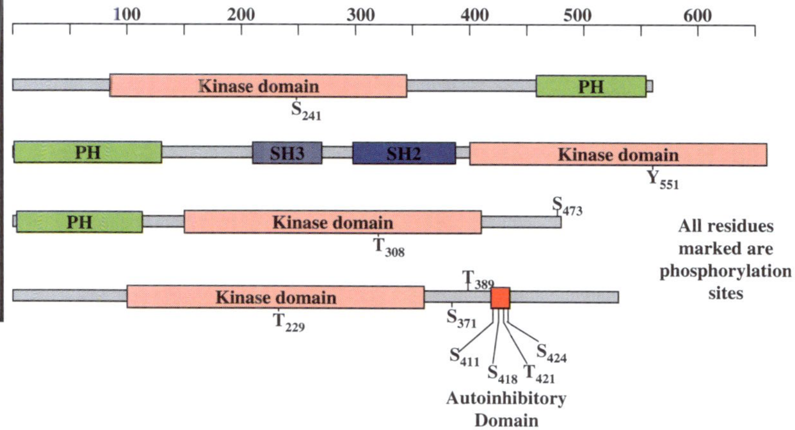

Information from UniProtKB/Swiss-Prot (http://us.expasy.org/uniprot) and references mentioned in the text. Some interesting points to note: (1) Variable location and size of PH domains relative to the other domains of the protein. (2) Presence of other domains (SH3 and SH2) in Btk allowing it to integrate multiple signalling pathways. (3) Key phosphorylation residues (serine, tyrosine and threonine) in the midst of the kinase domains. (4) Absence of a PH domain in S6K as its regulation is through intermediate kinases, not directly binding the product of the enzyme, inositol lipids phosphorylated on the D3 position.

Fig. 2 Diagrams of some PI3K regulated proteins.

PKB/Akt substrate is the forkhead family of transcription factors. The activity of this pathway is kept in check by the lipid phosphatase, PTEN. This conserved pathway has gained more members, and probably more functions, through evolution. In *C. elegans*, PI3K regulates differentiation and life span. In *D. melanogaster*, a role for PI3K in the regulation of cell size has been identified. During the evolution of the immune system, PI3K has developed a key role in antigen receptor signalling but its role in cell survival, proliferation and differentiation has also been maintained. It has also been shown to play a role in cell migration, particularly in response to chemokines. In cancer many of these cell processes are altered, so the role of these molecules in malignancy and how they might be targeted for therapeutic benefit are also described.

PI3K can be activated by a variety of cell receptors.[3–5] The most conserved is the activation of PI3K by receptors through tyrosine phosphorylation. Receptors that activate PI3K can have integral tyrosine kinase domains such as the platelet-derived growth factor receptor, or may bind or recruit tyrosine kinases. The interleukin-2 receptor, for example, is made up of three polypeptide chains, one of which binds a tyrosine kinase Jak1 whilst another binds the tyrosine kinase Jak3 (Fig. 3).[6] These kinases phosphorylate each other, the IL-2 receptor and various adaptor molecules, which in turn can recruit PI3K to the cell membrane leading to activation. Activation of the immune system depends on the antigen receptor of which there are three major types: the B-cell receptor, the T-cell receptor and the Fc receptors. As mentioned above, these receptors use PI3K activity to produce D3 lipids within the context of the receptor to promote the formation of antigen receptor signalosome. They do so via the recruitment of cytoplasmic tyrosine kinases, for example, the kinase Zap-70.

1.1. PKB/Akt

The best studied family of PI3K regulated molecules are the PKB/Akt kinase family.[3] PKB/Akt was identified as a protein kinase with high homology to protein kinase A and C, and thus gained the name PKB and a name RAC (Related to A and C).[7] RAC is no longer used because of confusion with a G-protein of the same name. It was also identified as the cellular homologue of the viral oncoprotein v-Akt, and thus gained the name Akt. The PKB/Akt family has three mammalian isoforms encoded by separate genes named by Human Genome Organisation (HUGO; www.gene.ucl.ac.uk/nomenclature/) as AKT1, AKT2 and AKT3. The proteins generated are closely related and are entitled PKBα, PKBβ and PKCγ, respectively. PKB/Akt is conserved in *C. elegans*, the slime mould, *Dictyostelium* and *D. melanogaster*.

PKB/Akt is a serine/threonine kinase of 57 kDa. It has three domains: a PH domain at the N-terminus, a kinase domain in the middle of the protein and regulatory domain at the C-terminus. The PH domain has high affinity for $PI_{(3,4)}P_2$ and $PI_{(3,4,5)}P_3$. Some elegant studies have used this domain as a sensor for PI3K activity in cells and transgenic mice and have shown that PI3K is rapidly produced *in vivo* following antigen receptor activation.[8–10]

On the basis of the kinase domain, PKB/Akt has been placed in the AGC protein kinase extended family.[11,12] AGC stands for cAMP-dependent protein kinase (PK**A**)/protein kinase **G**/protein kinase **C** (PKC). All members of this family display a high degree of primary sequence conservation in their kinase domains. Outside the catalytic domains, there is less similarity.

Following PI3K activation, the PH domain of PKB/Akt binds D3 lipids and becomes transiently associated with the inner surface of the cell membrane. This transient interaction results in phosphorylation of the kinase at two sites (Fig. 3).[3] Both are required for PKB/Akt activation. In PKBα/Akt1, these sites are threonine 308 and serine 473. PDK1 has been shown to phosphorylate PKB/Akt at Thr 308 in a $PI_{(3,4)}P_2$- or $PI_{(3,4,5)}P_3$-dependent fashion. Phosphorylation of PKB/Akt at serine 473 was observed first in 1997 but remains the subject of much investigation.[13] While many different kinases have been proposed, recent work has proposed a complex between the mTOR protein kinase with Rictor:GβL with strategies to inhibit mTOR or Rictor causing ablation of serine 473 phosphorylation in cells.[14]

After phosphorylation, there is evidence that PKB/Akt is released from the membrane and moves to phosphorylate various substrates throughout the cell (Fig. 3). These substrates include the FoxO family of transcription factors, the kinase GSK-3 and Bad, a molecule involved in regulating apoptosis (Fig. 4).

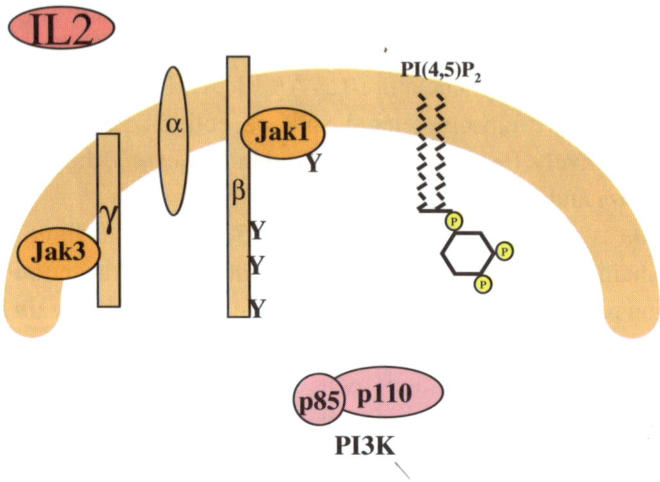

Fig. 3a Unstimulated cells. No tyrosine (Y) phosphorylation of the receptor. PI3K is not located at the membrane. $PI_{(3,4,5)}P_3$ is low.

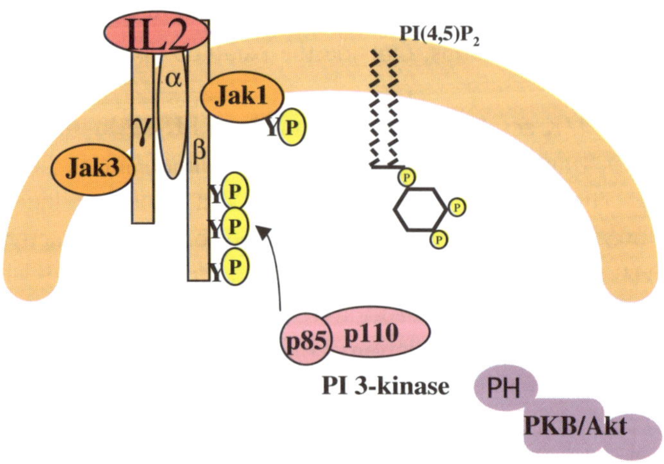

Fig. 3b Following receptor oligomerisation, tyrsoines in the receptor and the tyrosine kinases become phosphorylated leading to the generation of docking sites for the SH2 domain of the p85 subunit of PI3K. PKB/Akt and PDK1 are located in the cytosol.

1.2. PDK-1

PDK-1 (HUGO ID = PDPK1) was named as **p**hospholipid **d**ependent **k**inase-1 as the *in vitro* activity of the kinase was increased following the addition of phospholipids.[3,15,16] It was purified as an enzyme that could

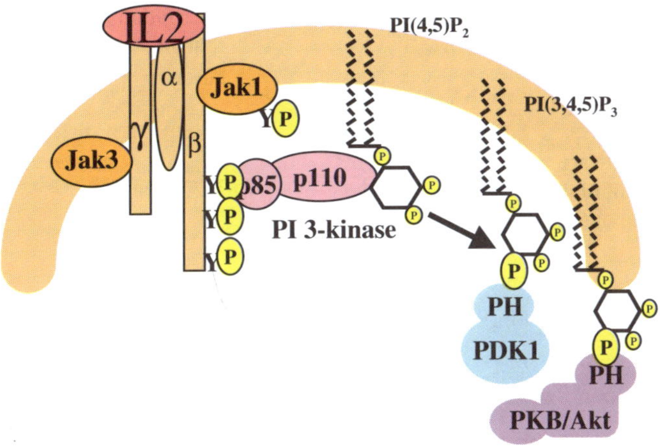

Fig. 3c PI3K, located at the cell membrane results in the production of PI(3,4,5)P3 from PI(4,5)P2. This provides docking sites, at the membrane for PDK1 and PKB/Akt bringing these molecules close together.

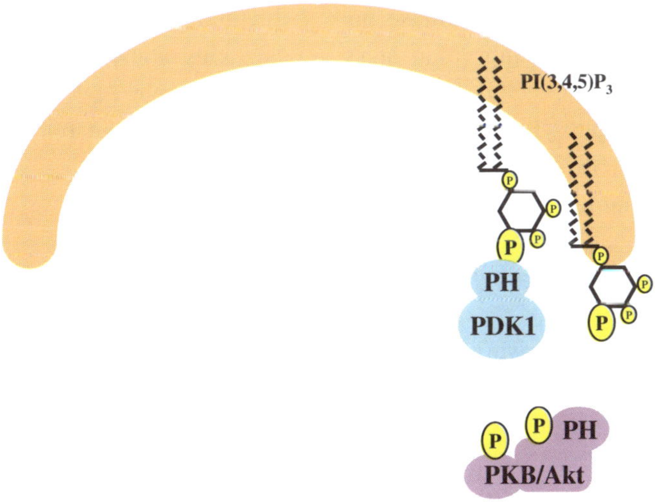

Fig. 3d PKB/Akt is phosphorylated on at least two sites, is released from the cell membrane to phosphorylate its downstream targets.

phosphorylate serine 308 of PKBα. It is a protein of 556 amino acids with a kinase domain at its N-terminus (amino acids 7–359) and a PH domain at its C-terminus, and has a molecular weight of 63 kDa. The PH domain has a high binding affinity for inositol phospholipids phosphorylated on

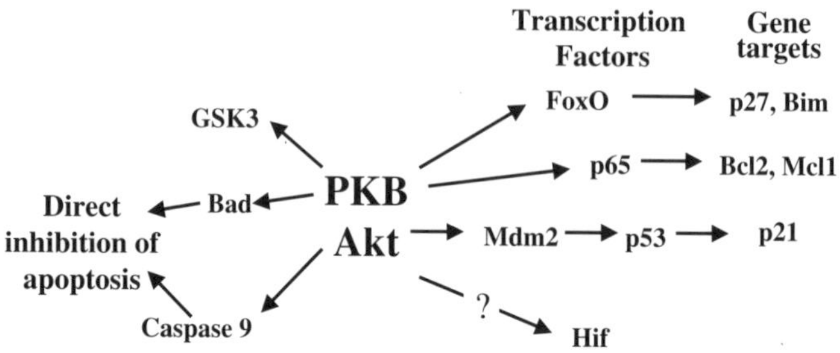

Fig. 4 Summary of some PKB/Akt targets.

the D3 position of the head group. PDK1 is well conserved through evolution with orthologs in mouse, *C. elegans*, *D. melanogaster* and *S. cerevisiae*.[3]

As indicated by the method of purification, PDK-1 plays an important role in the activation of PKB/Akt. However, PDK-1 also phosphorylates isoforms of S6 kinase and serum and glucocorticoid-inducible kinase (SGK).[11] Embryonic stem cells lacking PDK1 fail to activate PKBα, S6K1 and SGK1.[17,18] There is evidence that PDK1 in involved in the phosphorylation of all isoforms of PKB/Akt and p90 ribosomal S6 kinase (RSK). All of these proteins fall into the AGC protein kinase extended family. Many of this family of proteins contain consensus PDK1 phosphorylation sites. For this reason PDK1 is sometimes referred to a key or 'master' regulator of many of the kinases in this extended family.[11]

Somewhat surprisingly, the catalytic activity of PDK1 does not appear to be directly regulated since it can be immunoprecipitated from stimulated or unstimulated cells with the same high catalytic activity.[11] The binding of D3 lipids to its PH domain does not appear to alter the activity either. It is phosphorylated at serine 241 but this can be observed in bacterially expressed forms of the protein and there is evidence that this is a trans phosphorylation catalysed by other PDK1 interacting kinases. Thus, a key aspect of PDK1 regulation of PKB/Akt occurs when the molecules are brought together at the cell membrane following PI3K activation and production of D3 phosphorylated phosphotidylinositol.

1.3. Tec kinases and phospholipase C

PI3K activity is important for the functioning of the immune system.[5,19] In immune cells, $PI_{(3,4,5)}P_3$ plays a complex role in the formation of the 'signalosome'. The proper formation of this multiprotein signalling complex requires the presence of $PI_{(3,4,5)}P_3$ and specific adaptor proteins. Proteins containing PH domains with high affinity for $PI_{(3,4,5)}P_3$, such as phospholipase Cγ and members of the Tec family of kinases, Btk, Itk and Tec, are recruited to the signalosome via their interaction with $PI_{(3,4,5)}P_3$. The proper formation of the signalosome is required for antigen receptor signalling, culminating in an increase in internal Ca^{2+} levels that is reduced when PI3K activity is abrogated. A lower Ca^{2+} flux, in turn, reduces lymphocyte activation, cytokine production and proliferation.

1.3.1. *Molecular details of Tec family*

The Tec kinase family has five mammalian members, Btk, Bmx, Itk, Tec and Txk which are primarily expressed in the haematopoietic system.[20,21] The family is defined by a common structure that includes a tyrosine kinase catalytic domain occupying the C-terminal half of the protein preceded by a SH2 and an SH3 domain. They also contain a proline-rich domain that may play a role in regulation of the kinases. Three of the five members of the family also contain a PH domain. Two of these kinases, Btk and Itk, play an important role in B and T-cell signalling, respectively. The PH domain of the Btk and Itk is essential for the localisation of these kinases to the lipid raft area of the plasma membrane following antigen receptor signalling.

Btk or Bruton's tyrosine kinase is a protein of 659 amino acids and is 76 kDa in size. To integrate into the signalsome complex, Btk requires the presence of $PI_{(3,4,5)}P_3$ to which the PH domain will bind. At the same time, the SH2 domain of the protein recognises a phosphorylated tyrosine in the scaffolding protein BLNK/SLP-65. Two tyrosine phosphorylation sites have been identified. The phosphorylation of tyrosine 551 of Btk probably occurs after recruitment into the signalosome and regulates the transition between an active and inactive kinase. Btk can then act on its substrates to result in the activation of PLCγ, and the subsequent generation of a Ca^{2+} flux leading to B-cell activation. However, this somewhat simplistic scheme of interplay between PI3K and Btk has been challenged recently with a study in mouse cells suggesting that Btk can be fully activated in the absence of PI3K but that there is cooperation between the pathways further downstream.[22]

The importance of the PH domain to Btk is illustrated by mice with X-linked immunodeficiency (Xid) and patients with X-linked agammaglobulinaemia.[23–25] Mutations in the PH domain of Btk cause a dramatic reduction in the cellular response to ligands through this domain. One particular form of Xid is caused by a single point mutation (R28C) in the PH domain of Btk that abolishes selective binding to $PI_{(3,4,5)}P_3$. This loss of binding prevents the proper activation of the kinase and thus B-cell activation.

Itk is regulated in a similar way to Btk but is expressed in T-cells.[21] Human Itk is a protein of 620 amino acids and is 72 kDa in size. It has a similar structure to Btk with PH, proline rich, SH3 and SH2 domains preceding the kinase domain in the C-terminus. Itk-deficient mice have specific defects in the development of T-cell subsets. Some of the functions of Itk are likely to be due to its interaction with vav, a nucleotide exchange factor. Vav activates the Rho family of small GTPases, rac, rho and cdc42 which are important for actin reorganisation and for lymphocyte function.[26,27] There is a role for PI3K in actin reorganisation but this will not be discussed here.

1.3.2. *Molecular details of PLC*

Phospholipase C (PLC) describes a family of enzymes that catalyse the hydrolysis of phosphatidylinositol-4,5-bisphosphate into diacylglycerol and inositol-1,4,5-trisphosphate.[28,29] There are four sub-families of enzymes, PLC β, γ, δ and ε. PLC β, γ and δ all contain PH domains that are capable of binding $PI_{(3,4,5)}P_3$ but may also bind other phosphatidylinositides.[29] PLC-β has been shown to be recruited to the membrane following G-protein coupled receptor activation while PLC-γ is regulated by receptor protein tyrosine kinases, such as PDGF, and by antigen receptors, such as the T-cell receptor. For both PLC-β and PLC-γ the recruitment depends, at least in part, on the ability to bind PI3K products.

The antigen receptor activation of PLCγ has been extensively studied.[29,30] PLC-γ1 lies downstream of the T-cell receptor while PLC-γ2 lies downstream of the B-cell receptor. PLC-γ2 is described in a little more detail. PLC-γ2 contains a PH domain that plays a role in recruiting the enzyme to the signalosome. Like that of Btk, the PH domain of PLCγ2 also has high affinity for $PI_{(3,4,5)}P_3$. This domain functions in a variety of ways. Firstly, the binding of $PI_{(3,4,5)}P_3$ helps to recruit PKCγ2 to the signalosome. Secondly, this brings it close to the kinases, such as Btk, that phosphorylate PLCγ2 to increase its enzymatic activity. Thirdly, the membrane localisation of PLCγ2 juxtaposes it with its phospholipid substrate, in order to catalyse the generation of IP_3 and diacylglycerol (DAG). IP_3 is essential for increasing intracellular calcium, a key signalling mechanism to trigger lymphocyte activation and thus the immune system as a whole. PLCγ1 has a similar structure and probably plays a broadly similar role in the T-cell receptor synapse. Thus, PI3K and its products, the D3 lipids, play an essential role in the regulation of both lymphocyte activation through the formation of this signalosome.

2. Genetic Studies of PI3K Function in Model Organisms

Studies in model organisms, either deleting or over-expressing subunits of PI3K and other molecules in the pathway, are extremely powerful for dissecting PI3K function *in vivo*. Various model systems have been studied. Progressing through the animal kingdom, PI3K subunits have been identified and studied in the worm, *Caenorhabditis elegans*, the fruit fly, *Drosophila melanogaster*, and mice as well as mammalian cell cultures systems. Model organisms allow genetic analysis whereby genes can be deleted or over-expressed and the phenotype within the whole animal can be studied. All of the models investigated have shown that PI3K plays an important role in cell survival, cell size and cell proliferation and differentiation. Genetic analysis has also allowed an investigation of the roles of the PH domain containing molecules described above and has identified conserved pathways of molecules. However, phenotypes are not always as simple as we would hope.

2.1. Studies in *C. elegans*

In *C. elegans*, homologues for an insulin receptor (Daf-2), PI3K (Age-1) PDK-1 and Akt (Akt-1 & Akt-2) have been described.[31] Using genetic and biochemical studies, these have been assembled into a pathway, which when over-active causes extended life span for the worm. This pathway ends with Daf-16, which is a homologue of a forkhead transcription factor. Forkhead transcription factors are discussed in more detail in the next chapter. The production of D3-lipids, and thus the whole pathway, is regulated by a PTEN homologue called Daf-18. The effects on life span and dauer formation have demonstrated a role for this pathway in cell survival and differentiation.

2.2. Studies in *Drosophila*

Genetic studies in Drosophila suggests that the PI3K, and downstream effector molecules, play a key role on the regulation of cell size.[32,33] Null mutations of PI3K and PKB/Akt are lethal, but weak loss of function of dAkt1 produces small flies that have small cells. In agreement with this are experiments showing that over-expression of Dp110 and dAKT1 results in larger cells but has only a mild or no effect on cell number. Drosophila p70S6kinase, DS6K, lies downstream of PI3K and genetic studies shows very similar observations, with loss of DS6K resulting in slow imaginal disc growth leading to smaller flies with smaller cells.

Chapter 2 of this book details the different classes of PI3K. The number of proteins in the different classes and the number of adaptor subunits of PI3K has increased through evolution. A recent study in *D. melanogaster* has identified a distinct role for the Class II PI3K family as compared to the Class I family.[34] This study showed that while Class I PI3K regulated cell size in the fly, Class II regulated patterning. Over-expression of the class I PI3K (PI3K_92E.Dp110) generated larger organs (and other adult structures) without affecting patterning. However, over-expression of the Class II PI3K (PI3K_68D) altered wing vein patterns and sensory bristles. This suggests that during evolution different classes of PI3K have developed different functions.

2.3. Studies in mice

2.3.1. *PI3K*

The transition from lower to higher eukaryotic species is also associated with an increased number of PI3K subunits. There is also an increase in the number of different cell types and the complexity of these organisms. One important development, subsequent to the common ancestor of insects and mammals, has been the evolution of an adaptive immune response. This immune response has the capacity to 'adapt' to a pathogen to generate a specific response and to 'remember' this response in case the organism subsequently encounters such a pathogen again. Within the context of the immune system, PI3K has developed a number of specialised functions. The immune system also has the capacity to contribute to the pathology of many conditions, for example disregulated inflammation, autoimmune disease and asthma. PI3K plays a role in the activation of the immune system and in directing cells of the immune system to the site of disease. For this reason, PI3K therapeutics may have a relevance to many diseases.[35]

Gene targeting in mice has been valuable in studying PI3K and molecules downstream of the D3 lipids. Technology is available for the deletion and expression of specific genes in specific tissues. Also, a technique called rag complementation allows the study of molecules that are required for normal mammalian development to be studied in the context of the immune system. Simplistically, this system involves using genetically modified cells to reconstitute the immune system in mice in which the immune system in lacking. Thus genetically altered cells are surrounded by a 'normal' or wild type body. These various systems have been used to study the role of PI3K in the immune system and other tissues.

2.3.2. *Deletion of PI3K subunits*

A vast amount of data has been generated by the analysis of gene-targeted mice. These have been reviewed in detail in a number of publications.[19,36,37] In this chapter, only a flavor of the studies regarding the metabolic and immune cell phenotypes that have been observed is given. In some of the mutant mice, alterations in expression of one PI3K subunit can change the expression of other members. This is particularly the case for the deletion of the adaptor subunits. For this reason, some care must be taken in allocating function to particular molecules.

The mouse p85 locus encodes three distinct adaptor subunits by alternative splicing, p85α, p55α and p50α. Different gene deletion strategies targeting this locus have been performed. Deletion of all three forms results in mice which die shortly after birth.[38] Mice which lack only p85α, but not p55α and p50α, are viable but are hypoglycaemic and show reduced insulin and IGF-1 signalling.[39] These mice are hypoglycaemic and show reduced insulin and IGF-1 signalling. These mice also have reduced number of B-cells and reduced B-cell proliferation following antigen receptor stimulation.[22] Cytokine-induced proliferation of mast cells is also inhibited but FcεRI degranulation, another antigen receptor-mediated event, is not affected. T-cell proliferation is not affected either.

The mice lacking just p85α but retaining p55α and p50α have a more subtle phenotype.[22,39] They show increased insulin sensitivity and increased PI_3P production following insulin stimulation. This suggests that rather than just being involved in recruitment and activation of PI3K to receptors, p85α is somehow responsible for the negative regulation of PI3K activity. No mechanism has been proposed for this yet. These animals also display reduced B-cell proliferation[22] and altered T-cell development, probably due to an effect on dendritic cells which have increased IL-12 production which is a known regulator of T-cell development.[40]

Mice lacking the p85β adaptor subunit are viable, have similar insulin sensitivity to mice lacking p85α but have no B-cell phenotype. T-cells show increased T-cell proliferation. This supports the hypothesis that the immune system has developed specific roles for PI3K and suggests that the negative regulation by p85α may also occur for p85β.

The genes encoding p110α and p110β have also been deleted but mice deficient in these molecules die as embryos or shortly after implantation.[41,42] This shows the importance of these molecules to development but make further study difficult. Specific deletion of these genes in tissues is currently under development in a number of laboratories. Transgenic mice expressing p110α in the heart exhibited increased heart and cardiomyocyte size, while expression of a kinase dead version reduced heart size.[43] This indicates that the regulation of cell size by p110α is similar to that seen in other organisms and most likely represents an important phenotype for this subunit of PI3K.

The catalytic subunits p110δ and p110γ have a more restricted tissue distribution, being found in mainly in leukocytes. Mice lacking these genes survive without any gross abnormalities.[19,36,37] Strategies to destroy the catalytic activity but leave the protein expressed have also been successful.[44]

Only the study of p110γ has reported a metabolic activity which was the reduction of glucose stimulated insulin secretion. The immune cell phenotypes of both these mutants have been the most interesting.

B-cells from p110$\delta^{-/-}$ and p110δ kinase dead mice show reduced B-cell proliferation and consequent reduced antibody production. T-cell proliferation is reduced and mast cell degranulation is inhibited.[44–46] This suggests that antigen receptor signalling requires p110δ catalytic activity. The *in vivo* allergic response, which involves mast cells was inhibited showing a role for PI3K in allergy.[47]

Mice lacking p110γ showed distinct phenotypes demonstrating some unique functions of the two catalytic subunits despite similarly restricted expression.[48–50] The proliferation of B-cells from p110$\gamma^{-/-}$ mice were not affected. Chemotaxis of neutrophils, macrophages and dendritic cells was inhibited in both p110$\gamma^{-/-}$ mice and the mice lacking catalytic activity of p110γ. Additionally, T-cell proliferation was also inhibited in p110$\gamma^{-/-}$ mice.

A simplistic explanation is that p110δ and p85α are critical for B-cells, while T-cells may require both p110δ and p110γ. The double knockout of p110δ and p110γ causes a block in T-cell development and renders peripheral T-cells abnormal.[51] The role of PI3K in cell movement is important for the innate immune cells such as neutrophils and macrophage, a phenotype in which p110γ is critical. The p110α and p110β subunits are probably involved in the regulation of cell size, survival, proliferation and development at a fundamental level and so no viable organism can be produced in their absence. This illustrates a key concept that the core function for PI3K, mediated by p110α and p110β, evolved to become more cell type specific during the development of the innate and adaptive immune system, for example mediated by p110γ and p110δ.

2.3.3. *Alteration of PDK1*

Deletion of PDK1 is also lethal with PDK1$^{-/-}$ embryos dying at embryonic day 9.5 with multiple abnormalities.[52] Mice have been generated with reduced PDK1 expression and these mice are approximately half the size of wild type mice. Fibroblasts have been generated from the mice with reduced PDK-1 expression. These cells have a reduced size but no defect in proliferation, demonstrating a critical role for PDK1 in the regulation of cell size.

Two tissue-specific knockouts of PDK1 have been generated allowing further study of PDK1 *in vivo*. One of these analysed PDK1 function in the thymus, the organ of T-cell development where cells undergo ordered differentiation and cycle of proliferation to generate functional helper and cytotoxic T-cells.[53] Complete PDK1 loss blocked T-cell differentiation in the thymus, whereas reduced PDK1 expression allowed T-cell differentiation but blocked proliferative expansion.

Similar technology was used to generate mice lacking PDK1 in the liver.[54] These mice died at between four and 16 weeks of age due to liver failure, although the exact mechanism is not entirely clear. However, following administration of insulin blood glucose levels failed to return to the normal levels thereby demonstrating a role for PDK1 in metabolism. There was a loss of PKB/Akt activation in the liver following injection of insulin *in vivo*. There was reduced induction of gene expression by insulin. This study shows a clear role for PDK1 in the regulation of liver metabolism.

To summarise, PDK1-deficient mice show that like PI3K, PDK1 plays important roles in metabolism, regulation of cell size and has a role in the immune system, particularly it is required for T-cell development and proliferation.

2.3.4. *Alteration of PKB/Akt in vivo*

Many studies have shown that PKB/Akt can increase cell survival *in vitro*.[55,56] These have used constitutively active mutants of PKB/Akt, sometimes a mutant consisting of a fusion between Akt and the viral protein, gag. This mutant form of PKB/Akt has been expressed in T-cells using a specific T-cell transgenic promoter[57,58] and constitutively active forms of PKB/Akt have been expressed in other tissue.[55] T-cells containing constitutively active Akt show increased survival in response to apoptosis stimuli, induced via Fas ligand stimulation. Whilst this does not completely inhibit apoptosis, it increases survival in approximately 15% of the number of cells. While it sounds as if this may not be significant, spontaneous cell death is also inhibited with approximately ten-fold more cells surviving eight days after removal from the whole animal. This shows that expression of the gagPKB mutant increases the apoptotic threshold of these cells. When mice were observed for between four and eight months, the number of T-cells in the spleen and lymph node were approximately four- and ten-fold higher, respectively. These T-cells expressed higher levels of CD69, an activation marker, and mice also showed an increased number of B-cells expressing CD23. There was spontaneous lymphocyte infiltration into the liver and mice from one study developed thymic lymphomas.[59] Together all this data indicates that PKB/Akt plays an important role in T-cell homeostasis and function *in vivo*.

Of course, PKB/Akt does not only play a role in T-cells. Similar to both PI3K and PDK-1, PKB/Akt has also been shown to alter organ size when expressed *in vivo*. Active mutants of PKB/Akt have been expressed in the heart, mammary gland, pancreas and prostate.[55] The over-expression of active PKB/Akt *in vivo* generally causes an increase in cell size, cell number and a reduction of cellular apoptosis.

One study allows an *in vivo* comparison with PI3K. An active or a kinase-deficient PKB/Akt molecule was expressed specifically in the heart.[43,60] The expression of the active PKB/Akt increased the size of the heart and the size of individual myocytes. Expression of the kinase-deficient PKB/Akt had little effect alone but inhibited the enlarging effect on the heart that occurs following expression of constitutively active PI3K *in vivo*. These phenotypes establish a genetic link between PI3K and PKB/Akt in the mammal. The generation of mice deficient in isoforms of PKB/Akt have greatly added to our knowledge of the role of this molecule in the mammalian body.[55] They have revealed, unexpectedly and for the first time, organ- and isoform-specific roles for the molecule.

One of the first reports detailed the generation of mice lacking PKBβ/Akt2, which appeared to be enriched in insulin-responsive tissues. The initial and subsequent studies demonstrated impaired insulin responses and glucose uptake into tissues.[61,62] This insulin resistance can progress to a severe from of diabetes. Furthermore, a mild growth deficiency has also been reported.

The deletion of PKBα/Akt1 revealed a distinct phenotype. The mice are smaller than wild type and more sensitive to genotoxic stress.[61,63] The reduced size is due, in part, to a role for PKBα/Akt1 in the regulation of placental development and foetal growth.[64] Interestingly, no defects in glucose metabolism have been observed showing different roles for the two isoforms. It appears that one of the causes of the different functions may be the ability to regulate translocation of glucose transporter 4, which is important for the role of PKBβ/Akt2 on metabolism.[65] PKBα/Akt1$^{-/-}$ mice also show a defect in platelet aggregation which results in increased bleeding.[66] It is anticipated that phenotypes will be identified in specific tissues for PKB/Akt isoforms.

The deletion of PKBγ/Akt3 also shows a distinct phenotype.[67,68] This isoform is essential for brain development, and mice lacking PKBγ/Akt3 have brains between 15% and 25% smaller. This may reflect a function for PKB/Akt acquired during evolution as the brain became larger in more complex organisms. PKBγ/Akt3 has a more restricted expression than the other isoforms with varied expression even within regions of the brain. It will be interesting to see if the effects on brain size can alter learning or behaviour.

Double knockout of both PKBα/Akt1 and PKBβ/Akt2 cause a severe growth defect in the mice.[69] They die shortly after birth. The cells exhibit a proliferative defect and particular defects in muscle size and bone development. The increased severity indicates that there is likely to be some cooperation between these two molecules. The fact that the mice can be generated at all may suggest some compensation of function of PKBγ/Akt3 or other pathways.

In haematopoietic cells, another kinase Pim-2 can compensate for some of the effects of loss of PKBα/Akt1.[70] By inducing the phosphorylation of 4EBP-1, and may be other pathways, Pim-2 allows Akt-1 deficient mice to maintain normal haematopoiesis. Understanding and identifying this kind of interplay may allow us to identify the compensatory pathways and thus develop rational drug target combinations for therapy of specific diseases.

2.4. Studies in human cells

Before the role of PI3K was studied in model systems *in vivo*, a large of body of work showed that PI3K played an important role in cell signalling, survival and proliferation. This work was performed in a variety of cell lines and used various strategies for altering PI3K activity in cells to study its function.

2.4.1. *Transient transfection of mutant molecules of PI3K*

One of the most important mutant molecules was a dominant negative form of p85, the adaptor subunit of PI3K. This was described by the Waterfield lab and it contains a mutation that prevents p85 from binding to the adaptor subunit p110.[71] Over-expression of this mutant is thought to bind and sequester phosphotyrosines that would otherwise recruit a functional PI3K and so results in a lower PI3K activity. Transient transfection of this mutant inhibits the proliferation and/or survival of a variety of cell types.[72–75]

Constitutively active mutants of the kinase domain of PI3K have also been generated. The strategies rely on localising the p110 catalytic subunit of PI3K to the cell membrane where it has access to its substrates.[76,77] This membrane localisation was sufficient to increase the production of PI3K substrates in various cell lines and often increases survival and proliferation where expressed. Furthermore, these active mutants were used to show that PI3K was sufficient for the activation of downstream molecules including PKB/Akt.[76,78] This approach has been very useful to separate phenotypes for which PI3K activity was sufficient, such as the activation of PKB/Akt.

2.4.2. *The pharmacological inhibitors, wortmannin and LY294002*

The development of two pharmacological inhibitors allowed the study of PI3K in cells that were not easily transfected such as primary cells.[79] Wortmannin is a hydrophobic steroid-related product of the fungus *Taloromyces* that potently inhibits PI3K signal transduction pathways in mammalian cells.[80] It irreversibly binds to the catalytic subunit of PI3K but has a short chemical half-life, and is unstable, particularly in aqueous solution. Vlahos *et al.* developed a structurally distinct PI3K inhibitor, LY294002, which is a competitive ATP-binding site inhibitor.[81] In contrast to wortmannin, it is a reversible inhibitor but is more stable than wortmannin in solution and has a longer half-life.

Both of these inhibitors have been used extensively to study the role of PI3K in cells with publications using these agents numbering in the thousands. The results observed vary according to cell type, doses and the systems used. However, both agents clearly show the importance of PI3K activity to a wide range of biological processes. Both agents can inhibit other members of the PIK superfamily. A group, led by Cohen in Dundee, have studied the specificity of both wortmannin and LY294002.[82] Wortmannin also inhibited SmMLCK while LY294002 also inhibited casein kinase-2 again showing the importance of caution regarding the interpretation of data obtained with the use of pharmacological inhibitors. Thus, it is possible that some of the phenotypes identified with these two drugs could be the result of inhibition of kinases other than PI3K. Consequently, it is important to use as many different strategies as possible to prove a hypothesis, including the use of multiple inhibitors, preferably of different structures and, if possible, to complement studies with pharmacological inhibitors with other strategies for inhibiting the same molecules, such as the mutant molecules described above.

Despite that cautionary note, LY294002 and wortmannin remain useful agents for the analysis of PI3K. They have been used to implicate PI3K in cell survival and proliferation, both phenotypes which have been shown in a wide variety of systems and using a variety of approaches including some of the model systems described above.

2.4.3. *Antibodies injected into cells*

An alternative approach to the use of inhibitors was the microinjection of specific antibodies into cells. This approach works particularly well with cells cultured on coverslips. It was used initially to study the role of specific isoforms of PI3K. However, it can be quite labour-intensive and has both the advantage and disadvantage that small numbers of cells can be intensively studied. This technique was used to show that p110α was involved in platelet-derived growth factor signalling.[83] It was also used to separate the role of p110α from that of p110β and p110δ in the regulation of mitogenesis and cell migration in macrophages.[84]

3. Molecules Downstream of PI3K and D3 Lipids

As mentioned previously, all of the signals downstream of PI3K activity must come from molecules that are altered in the presence of phosphorylated D3 lipids or other lipid products generated from phosphorylated D3 lipids. For this reason, they must lie downstream of some of the molecules altered by PI3K activity, such as the PH domain-containing molecules previously described. There is cross-talk between PI3K regulated pathways and other signal transduction pathways and, as mentioned for PLCγ, some molecules have the capacity to integrate signals generated by PI3K and protein kinases.

3.1. PKB substrates

3.1.1. *GSK-3*

The first PKB/Akt substrate identified was glycogen synthase kinase-3 (GSK-3).[85] GSK-3 has been shown to be altered is response to PI3K activity is a variety of systems.[86–88] The phosphorylation by PKB/Akt of GSK-3 inhibits GSK-3 activity. GSK-3 itself phosphorylates a number of other molecules including metabolic enzymes, for example glycogen synthase, and transcription factors, for example NFAT. The metabolic effects of PKB may result in part from its regulation of GSK-3. In the case of NFAT, the phosphorylation of NFAT by kinase such as GSK-3 drives nuclear export of the protein, and thus represses its transcriptional activity. The full interplay between all these molecules, and the relative importance in PKB function is still under investigation.

3.1.2. *FoxO transcription factor family*

Three members of this family have been shown to be directly phosphorylated by PKB/Akt in mammalian cells, and since then the FoxO family has been shown to be involved in both cell proliferation and survival.[89] Some of the first observations were made in *C. elegans* where a pathway from the insulin receptor through PI3K to PKB/Akt has been shown to suppress the *C. elegans* member of the FoxO family, called DAF16.[90] As a transcription factor, it has been implicated in the regulation of the p27kip1 promoter and the Bim promoter which in part explains its effects on cell survival and proliferation. The FoxO family is described in detail in the next chapter.

3.1.3. *Bad and caspase-9*

Bad forms a heterodimer with Bcl-2 or Bcl-X_L, two anti-apoptotic proteins.[88] By making this dimer, Bad represses the anti-apoptotic function of Bcl-2. When phosphorylated by PKB/Akt at serine 136, Bad can no longer bind Bcl-2 or Bcl-X_L, freeing these molecules to prevent apoptosis. This is one of the mechanisms by which PI3K and PKB/Akt can increase cell survival. Human caspase-9, but not mouse caspase-9, is another target for PKB/Akt which may also explain the effects of PKB/Akt on promoting cell survival.

3.1.4. *IκB kinase and interplay with the transcription factor, NF-κB*

Cross-talk between PI3K–PKB/Akt and the NF-κB transcription factor activation pathway has been proposed in a number of cell systems.[91–93] Over-expression of gagPKB, a constitutive active mutant of PKB/Akt, in mouse thymus, results in an increased activation of NF-κB following simulation of T-cells showing that interplay can occur *in vivo*.[58] Importantly, gagPKB could not activate NF-κB in these cells suggesting that active PKB/Akt alone is insufficient for NF-κB activation in contrast with other reports. NF-κB is involved in the expression of many survival genes such as Bcl-2 and Mcl-1. So the transcriptional regulation of NF-κB can also be an important factor in the pro-survival effects of the PI3K–PKB/Akt pathway. Various cell stimuli which activate NF-κB have been proposed to also activate PI3K, for example toll-like receptors, and the cytokines tumour necrosis factor and interleukin-1.[94] Some groups report an inhibition of the activation of NF-κB following treatment of the cells with PI3K inhibitors. However, this hypothesis is the subject of some debate. A mechanism resulting from an interaction with the RelA/p65 subunit of NF-κB has been proposed.[95] Generally, it seems that PI3K–PKB/Akt pathway may have the capacity to increase NF-κB transcriptional activity but is probably not able to activate NF-κB alone.

3.1.5. *Other targets for PKB/Akt*

PKB/Akt has been shown to activate Mdm2, which in turn regulates p53, a protein involved in the regulation of cell death in response to DNA damage. Also, PKB/Akt has been reported to phosphorylate Yap, a binding partner for the p53 family member, p73. All of these represent ways in which PKB/Akt can affect cell survival.[88]

There are also published papers showing that PKB/Akt can phosphorylate CREB[96] and a report identifying a transcription factor called trachealess in the fly.[97] One of the orthologs of trachealess in mammalian cells is the transcription factor hypoxia-inducible factor-1 (HIF-1),[98] and evidence has accumulated that this may be regulated by PKB/Akt,[99] although this is somewhat controversial. These targets, and others, are under investigation. Their relative importance are likely to be determined using model systems and in human cells over the next few years.

3.2. TOR, S6K and the regulation of protein translation

Growth factors increase translation to facilitate and enhance protein expression. PI3K activity has been shown to be required for this increase in translation and a complex pathway involving PI3K, PDK-1 and PKB/Akt has been identified as integrating with a molecule entitled TOR. TOR itself regulates at least two pathways, one which results in the phosphorylation of S6, a component of the ribosome, and another which alters the activity of translation elongation factors. Together, this network allows a dramatic increase in the activity of the protein synthesis machinery which is necessary for an increase in cell size and proliferation.

3.2.1. *Mammalian target of rapamycin*

Target of rapamycin (TOR) is a 289 kDa serine/threonine kinase and a member of the phosphoinositide-kinase related kinase family. It has been shown to be a central mediator of nutrient and growth factor signalling.[100] TOR is involved in the activation of S6Kinase (S6K) which is required for the phosphorylation of the 40s ribosomal subunit of S6, necessary to permit translation and cell cycle progression.[101] The S6K signalling pathway is activated by PI3K. A key event in the activation of S6K is the phosphorylation of the T389 residue of its linker domain and TOR is probably the critical kinase for this phosphorylation.[101] In the presence of rapamycin, the phosphorylation of S6K is prevented and this results in a prolonging of G1 transit times in the cell cycle and an inhibition of T-cell proliferation.[102]

Mammalian TOR (mTOR) was first identified and cloned shortly after the discovery of the two yeast genes TOR1 and TOR2 in the budding yeast *Saccharomyces cerevisiae*.[102] In addition to phosphorylation of S6K, mTOR also phosphorylates and inactivates 4E-BP1, a repressor of mRNA translation and a key factor in cell cycle regulation.[103] Mammalian genomes encode a single TOR protein with approximately 42% amino acid sequence homology to the yeast TOR proteins. The mTOR protein is 2549 amino acids in length and contains several distinct and conserved structural domains including a kinase domain at the C-terminus.[100] TOR is phosphorylated at serine 2481 via the PI3K pathway and is autophosphorylated via intrinsic serine/threonine kinase activity. In mammals, mTOR forms a complex with regulatory associated protein of TOR, raptor[104] and mLST8 and this complex is the mammalian equivalent of yeast TOR complex 1.[103] Evidence suggests that raptor functions as an adaptor protein leading to the recruitment of the mTOR substrates S6-kinase (S6K) and 4E-BP1.

3.2.2. *S6K and S6, one pathway regulating of translation*

S6K is another evolutionary conserved molecule.[32] There is one gene in Drosophila, the loss of which results in smaller flies. In mice and man, there are two genes called S6K1 and S6K2 on separate chromosomes (Hugo ID for S6K1 is RPS6KB1; Hugo ID for S6K2 is RPS6KB2). The gene for S6K1 can produce at least two proteins of different size. The best characterised is the protein entitled p70 S6K as it is approximately 70 kDa in size. Its activity is regulated in a PI3K-dependent fashion and by the activity of TOR (Target of Rapamycin). S6K contains at least six different phosphorylation sites that have been the subject of study for many years. The activation of S6K1 first requires the phosphorylation of its autoinhibitory domain by mTOR followed by phosphorylation of threonine 389 of S6K1 by mTOR. Threonine 389 phosphorylation allows PDK-1 to bind to S6K1. This binding permits PDK1 to phosphorylate S6K1 on threonine 229, resulting in S6K1 activation. This kind of hierarchy of phosphorylations can be seen in T-cells stimulated with the growth factor interleukin-2. IL-2 stimulates PI3K leading to S6K phosphorylation. This is inhibited by LY294002, the PI3K inhibitor. However, when rapamycin, the inhibitor of TOR is added, all of the S6K phosphorylations are lost even in the presence of an active PI3K-PDK1-PKB pathway.[105] S6K2 has analagous phosphorylation sites and is generally subject to a similar method of regulation to S6K1.[32]

One key substrate for S6K is the S6 ribosomal protein. Phosphorylation of the S6 ribosomal proteins is required for the translation of a set of

RNA molecules that encode molecules involved in protein translation. Deletion of S6 in mice causes a block in ribosome biogenesis and prevents cell cycle progression[106] showing a direct link to cell proliferation. S6 is phosphorylated at multiple sites and its phosphorylation is stimulated by mitogens. However, S6 can also be phosphorylated by kinases other than S6K.[107]

3.2.3. *4E-BP, another pathway regulating translation*

Another key mechanism for the increase in protein synthesis is the enhanced formation of a translation initiation complex that binds to the 5′ cap structure of eukaryotic mRNAs.[108,109] This is a multi-molecular complex entitled eIF4F. This complex is regulated by phosphorylation and a family of repressor proteins, entitled eIF4E-binding proteins, 4E-BP. When bound, 4E-BP prevents the formation of the eIF4F complex and the subsequent increase in translation.

Mitogen-induced phosphorylation of 4E-BP, prevents its interaction with eIF4F and thus results in an increase in translation.[108,110] TOR directly phosphorylates the 4E-BPs in a rapamycin-dependent fashion. Like S6K, 4E-BP1 contains multiple phosphorylation sites and phosphorylation occurs in a stepwise fashion. Phosphorylation at two sites, threonine 37 and 43, precede the phosphorylation at sites regulated by mitogens which include serine 65 and threonine 70. The phosphorylation of 4E-BP1 is stimulated by mitogens and is regulated, at least in part, by PI3K activity.

What seems apparent for TOR, S6K, S6 and 4E-BP, is that they are all capable of integrating signals from different pathways and potentially different cell stimuli. This integration involves phosphorylation and also involves the formation of multiprotein complexes. While the elucidation of these interactions is ongoing, the importance of these molecules to cell size, cell proliferation and malignancy is undeniable. Furthermore, the sensitivity of these pathways to rapamycin means they can be targeted in humans for the treatment of cancer and to facilitate transplantation, as discussed below.

4. PI3K Regulation of Cell Survival

As mentioned in the sections dealing with targeting PI3K and PKB/Akt function, *in vitro* and *in vivo*, there is now a clear consensus that the

PI3K, PKB/Akt pathway plays an important role in cell survival. Increasing PI3K and PKB/Akt activity, for example through growth factors, or through over-expression, renders cells more resistant to apoptosis, i.e. programmed cell death.

Wortmannin or LY294002, the PI3K inhibitors, have been widely used to identify a role for PI3K activity in the regulation of cell survival. Some cells, such as mouse lymphocytes, will die following the addition of a PI3K inhibitor. Other cell types seem to require the addition of additional pro-apoptotic signals such as the over-expression of c-myc.[111] Increasing the expression of pro-survival genes like bcl-2 increases the apoptotic threshold of cells, and reduces the effects of PI3K inhibitors. One of the most powerful demonstrations of the role of PKB/Akt in the regulation of cell survival has been the over-expression of a constitutively active form of the protein *in vivo*.[55] In many tissues, agreeing with work on cell lines, PKB/Akt can make these cells more resistant to cell death.[88] Over-expression of active PKB/Akt does not mean that cells do not die but rather lengthens their life span.

It has proven more challenging to elucidate the relative importance of the PKB/Akt substrates in regulation of cell survival. This probably reflects some redundancy or cell type variability within the mechanism by which PKB/Akt can alter cell survival. There are probably at least three distinct mechanisms by which PKB/Akt can alter cell survival.[88]

The first mechanism is a direct interaction with cellular apoptosis machinery. PKB/Akt has been shown to phosphorylate Bad, releasing bcl-2 to increase cell survival. PKB/Akt has also been proposed to phosphorylate human caspase-9. Together these represent two direct interactions with the apoptosis machinery.

The second mechanism is a change in the transcription of pro- or anti-apoptotic genes. PKB/Akt can change the expression of genes through the regulation of the FoxO family, NF-κB, Yap, or the p53 pathway. When gagPKB was over-expressed in the thymus, these cells showed an increase in NF-κB activity and an increase in Bcl-XL, a gene which NF-κB has been shown to transactivate. The forkhead family of transcription factors have the capacity to alter the expression of Bim, a topic discussed in more detail in the following chapter. The p53 pathway is involved in DNA damage checkpoint control. PKB/Akt has been shown to alter MDM2 phosphorylation, a molecule that alters p53 expression.

The third mechanism by which PKB/Akt may alter apoptosis is the alteration of energy provision through its effects on glucose metabolism through GSK3. Growing cells, in particular, require substantial energy for DNA synthesis and cell division. By promoting energy provision, PI3K and PKB/Akt may increase cell survival. This is discussed further below.

In vivo studies show that loss of PI3K can decrease lymphocyte proliferation.[37] This decrease in lymphocyte proliferation may be due to a decrease in cell survival. However, it may not only reflect direct regulation of cell survival genes but rather the lack of production of cytokines or their receptors that are required for cell survival. For example, lack of p110δ activity causes a reduction in intracellular calcium, probably because the synapse is not properly formed.[44] This reduction of intracellular calcium results in lower levels of cytokine, such as interleukin-2. Loss of cytokine production can contribute to a reduction in cell survival.

In summary, activation of PI3K and PKB/Akt can increase the apoptotic threshold of cells. Some cells are critically dependent on PI3K activity while others use it as one of many strategies to keep cells alive. PKB/Akt is a critical mediator of survival induced by PI3K. PI3K and PKB/Akt enhance cell survival by altering survival molecules such as Bad and also by changing gene expression to induce pro-survival genes such as Bcl-2 and cytokines, depending on cellular context.

5. Signalling from PI3K to Cell Proliferation

Notwithstanding the plethora of evidence that PI3K and its targets regulate cell size and cell proliferation, particularly in mammalian cells, it is important to try and link signalling molecules to the cell cycle progression. There is evidence that inhibition of PI3K prevents cells entering cell cycle. Treatment of mammalian cells with PI3K inhibitors prevents the expression of the D-type cyclins, for example cyclin D2 and cyclin D3 in T-cells.[112–116] Furthermore, the degradation of at least one member of the cyclin-dependent kinases, p27kip1, is decreased in the presence of the PI3K inhibitor, LY294002. This in turn prevents the phosphorylation of the pocket proteins and inhibits E2F transcriptional activity.

There is evidence for transcriptional and post-transcriptional regulation of the D-type cyclins. All three D-type cyclins contain complex gene promoters, which integrate multiple pathways. For example, cyclin D2 seems to require signals from the transcription factors STAT5, SP1 and CREB.[117,118] PI3K activity, while not sufficient, is required for maximal induction of these genes.[115,119]

The cyclin-dependent kinase inhibitor, p27kip1, is also regulated at multiple levels through PI3K regulated pathways. One important mechanism of transcriptional control is through the FoxO family of transcription factors, discussed in the next chapter. FoxO can bind to and regulate the p27kip1 promoter.[120,121] There is also a body of data showing that p27kip1 is regulated by ubiquitinylation and proteolysis. Rapamycin, which inhibits TOR, prevents p27kip1 degradation and thus cell cycle.[122,123] This shows a role for the PI3K/TOR pathway in post-transcriptional regulation of p27kip1.

PI3K has also been implicated in energy provision, particularly for the growing lymphocytes.[124] Lymphocytes undergo major size changes during activation and growth which requires energy, substantial gene expression and protein production. Lymphocytes treated with PI3K inhibitors remain small,[9] proliferate less and generate less ATP. Co-stimulation of T-cells, for example through the cell surface molecule CD28, has been proposed to stimulate PI3K and cause an increase in glucose transport into the cell and glycolysis.[125] PKB/Akt has been shown to increase the expression and cell surface localisation of Glut1, the major glucose transporter in lymphocytes. Thus by increasing the availability of ATP, PI3K–PKB/Akt increases the energy available for cell division and proliferation.

As mentioned before, the TOR S6K pathway is regulated by PI3K. TOR, in particular, plays a role in detecting nutrient supply, and both S6K and 4EBP-1 mediate PI3K pathways to translation.[100] Translation is essential for the increase in protein expression required for cell division. In order to facilitate cell division, the cell must increase the expression of many genes including those involved in synthesis of DNA and the maintenance of cellular structure and function. Without an increase in protein translation, this increase in gene expression would be of limited effect. Thus PI3K stimulates yet another pathway that facilitates cell division. It seems clear that PI3K and pathways downstream of it provide important signals that the mammalian cell integrates with other signals from diverse sources to facilitate cell proliferation.

6. Alterations of PI3K and Its Pathways in Cancer

PI3K was first identified as an enzymatic activity that was essential for the transforming activity of the V-src and polyoma middle T-antigen.[126] This immediately suggested that the kinase responsible for this activity would be involved in cancer. Subsequent studies have clearly demonstrated that PI3K activity and PI3K regulated pathways are altered in most cancers and that the role of PI3K in cell proliferation and survival make it a potentially exciting target for cancer therapy.[127]

PI3K activity is increased in tumours in a number of different ways. Firstly, molecules upstream of PI3K are often constitutively activated or over-expressed. This is the case for tyrosine kinase receptors, for example the Erb-B2 receptors, found in lung and breast cancer. The fusion proteins found in chronic myeloid leukaemia usually involve a part of a cytosolic tyrosine kinase, for example the Bcr-Abl fusion protein. Both the receptor and cytosolic tyrosine kinases normally cause an increase in PI3K activity. Cells that require the activity of these tyrosine kinases are usually dependent on endogenous PI3K activity. When treated with PI3K inhibitors, these cells either stop proliferating or die.

The small G-protein ras, particularly the mutated form found in cancers, has also been shown to increase PI3K activity. Mutations in ras are found in pancreatic, colorectal, lung and thyroid cancers among others.[128] Ras mutations in chronic myeloid leukaemia are probably involved in altering and even inhibiting cell differentiation. However, PI3K is not the only pathway downstream of ras or the tyrosine kinase mutations mentioned above. Thus, ras and the constitutively active tyrosine kinases cause increases in other cell signalling pathways too.

As described in detail in a previous chapter, PI3K activity is regulated by the lipid phosphatase, PTEN. In normal cells, PTEN appears to keep the levels of PI_3P tightly under control.[129] PTEN is mutated in glioma as well as prostate and endometrial cancers demonstrating the importance of this lipid for malignancy.[128,129] When PTEN is mutated, the basal levels of PI_3P lipids are higher and PI_3P levels are maintained longer than in wild type cells following cell stimulation. Mouse models of cancer have also demonstrated the importance of PI3K, PKB/Akt and PTEN. Targeted deletion of PTEN mimics loss of PTEN in human cancers and results in the development of tumours in lymphoid, thyroid, breast and other tissues. Over-expressing constitutively active PKB/Akt in T-cells in the mouse thymus can lead to thymoma.[59]

Current thinking about cancer is that it requires a number of genetic mutations in the many processes the cell uses to regulate survival, proliferation and differentiation. An increase of PI3K is involved in the majority of cancers and so altering this pathway is likely to be the result of one of these genetic mutations. For this reason, targeting this pathway for diagnosis or therapy is an ongoing area of research and clinical investigation.

6.1. Activation of PI3K by viruses

Viruses have been implicated in the pathogenesis of a number of human cancers. In some cases, this is due to the manipulation of cell survival and cell cycle by the virus. Examples of viruses that play a role in cancer are: human papilloma virus, which is a key factor in cervical cancer; Epstein-Barr virus (EBV), which has been implicated in Hodgkin's and Burkitt's lymphoma; and human T-cell leukaemia virus (HTLV-1), which contributes to adult T-cell leukaemia. To date, no clear role for PI3K has been implicated in papilloma virus pathogenesis, but both EBV and HTLV and other viruses, have the capacity to modulate PI3K.[130–132] In the case of EBV, viral gene products have been shown to activate PI3K and downstream pathways. There are also cytokines that induce PI3K, that have been shown to be induced by EBV. Thus, the virus may have developed multiple mechanisms to increase and maintain PI3K activity. The transformation of fibroblasts by HTLV-1 has been shown to require PI3K activity.[132] The mechanism for this has not yet been established but constitutive chemokine production has been shown in adult T-cell leukaemia cells. Chemokines have the capacity to activate PI3K.

7. Cancer Therapeutics

An evolutionary conserved pathway from tyrosine kinases, to PI3K, to PDK-1 and PKB/Akt has been previously described. This pathway has the capacity to regulate cell proliferation, cell survival and cell differentiation. As these are key molecular processes involved in cancer, targeting this pathway is seen as valuable for cancer therapy and is an area of ongoing cancer research and investigation by pharmacological companies. All of these and other downstream molecules are targets for the development of small molecules to alter their function.

7.1. Targeting tyrosine kinases

Many strategies for targeting tyrosine kinases have been developed. One major victory for this area of research is the development of inhibitors of Bcr-Abl kinase, found in most cases of chronic myeloid leukaemia. The drug is called Glivec or Geevec which was developed by Novartis (www.glivec.com). Patients treated with this drug enter remission, particularly when diagnosed and treated early on in the disease progression, but usually are not cured. Unfortunately, when the disease is advanced, further mutations can occur which render the kinase resistant to the inhibitor. Another good example is gefitinib (IRESSA) currently being marketed by AstraZeneca. The company have information at www.iressa.com. It is an orally active inhibitor of the epidermal growth factor tyrosine kinase which is being used to treat non-small cell lung cancer and breast cancer.[133] Both of these drugs inhibit PI3K activity in their relevant cancer tissue, or appropriate cell lines. Other methods, including monoclonal antibodies, are also being used to target receptor tyrosine kinases with some success.

7.2. Targeting PI3K itself

Two PI3K inhibitors, wortmannin and LY294002, have previously mentioned. While these are reasonably specific for the PI3K, they show very little subtype specificity for the kinases. As PI3K has been implicated in the survival of lots of different cell types, there is a concern that it will be difficult to utilise a PI3K inhibitor without causing significant side effects. Data from mice, where these inhibitors have been administered shows a high level of toxicity. Thus, one key factor for the development of anti-cancer agents targeting the PI3K pathway is the identification of specific inhibitors for individual types of PI3K.[79,126]

Some important developments have been made in this area. One of the most interesting agents currently available is a compound developed by

ICOS corporation (www.icos.com) which specifically targets p110γ, a catalytic subunit that is selectively expressed in leukocytes, cells of the innate and adaptive immune system. This agent will inhibit leukocyte migration and is probably going to be more useful for other conditions rather than cancer.[134] It may prove useful for cancers that affect leukocytes such as some leukaemias, particularly as PI3Kγ-deficient mice have altered proliferation (see above for details).

Currently at least three other companies have patents on what they show or claim to be specific PI3K inhibitors. These include PIramed (Patent no: US6403588 B1), Kinacia (Patent No: 0153266 A1) and Novartis (WO 03-72557 A1).[79] Most of these agents are similar in part to LY294002 in structure or contain similarities to caffeine, which has also been shown to inhibit PI3K and PI3K-like enzymes. These agents are being developed not only to treat cancer but also may play a role in the treatment of other conditions such as respiratory disease, rheumatoid arthritis and possibly hypertension.

There is also a lot of ongoing investigation into developing new agents. Two recent developments include an opioid antagonist, Naltrindole,[135] and a rotenoid, deguelin.[136] The agents are described as inhibitors of 'Akt/PKB signalling' or a 'PI3K/Akt inhibitor'. While they cause apoptosis of cancer cells, the detailed mechanisms of action is currently unknown. This area of research is ongoing and is likely to produce much excitement in the coming years.

7.3. Targeting PDK-1 and PKB/Akt

Both PDK-1 and PKB/Akt inhibitors are under investigation. There appears to be only one form of PDK-1 in the mammalian genome and PDK-1 has been shown to regulate the phosphorylation of a large number of kinases. This suggests that while specific inhibitors may be developed, similar toxicity might arise from the use of PDK-1 inhibitors *in vivo*. Some of the potential toxicity problems may be overcome by targeted delivery to tumours or other sites of interest. The team that originally described PDK-1 has also been making very interesting discoveries in the study of PDK-1 inhibition.[137,138] They have combined a structural approach with strong enzyme kinetics and have screened inhibitors against a range of kinases, in a method called kinase profiling. They have studied a molecule called UCN-01, which has been described as a PDK-1 inhibitor. However, their study showed that UCN-01 also inhibited more than ten other kinases, including protein kinase C and S6 kinase.[138] Thus while this agent is being evaluated in clinical trials it cannot be considered as a specific PDK-1 inhibitor. The same researchers have also studied crystals of PDK-1 with a variety of other small molecules, including a molecule entitled LY333531.[137] By this rational and diligent approach, it seems possible that specific inhibitors of PDK-1 can be developed.

The existence of three genes encoding isoforms of PKB/Akt and their specific functions elucidated by deletions of the genes in mice make them very exciting targets. A major problem for researchers developing PKB/Akt inhibitors is the similarity of the molecules to PKA and PKC. However, there has been much progress in developing inhibitors with specificity for PKA and PKC so previous experience suggests that PKB/Akt inhibitors may be identified. Similar to PI3K the role of PKB/Akt in regulation of survival of normal cells, not just malignant cells, may mean that even specific inhibitors may exhibit some unwanted toxicity. Also, while there are three isoforms, they are closely related, making the generation of isoform-specific inhibitors more difficult.

A screen of the Cancer Institute Diversity Set has identified a small molecule inhibitor of PKB/Akt.[139] This inhibitor, known as triciribine, was shown to inhibit PKB/Akt phosphorylation but does not affect PDK-1 or PI3K activity. While it inhibits the growth of PKB/Akt expressing tumours *in vivo*, the data suggests that it does not act as a an ATP or substrate competitor. Similar results have been observed with other inhibitors. This is an exciting development suggesting potential therapeutics can and will be delivered to the clinic.

7.4. Rapamycin and analogues

Rapamycin, discovered through analysis of soil samples isolated from Easter Island, is a lyphophilic macrolide produced by a strain of *Streptomyces hygroscopicus*.[40,141] It is also known as sirolimus or Rapamune and is marketed by Wyeth Pharmaceuticals (www.rapamune.com). The work of Calne and others established rapamycin's role in the clinic and this natural microbial product has potent immunosuppressive activity.[142] It recently gained approval for the prophylaxis treatment of renal transplant rejection as an adjunct to cyclosporine A (CsA), another immunosuppressant and corticosteroids. Unlike CsA, rapamycin is not associated with nephrotoxicity but administration to transplant patients is associated with other side-effects.[143]

Cyclosporine A was one of the first major immunosuppressive drugs. It inhibits the activation of the calcium-dependent phosphatase, calcineurin and thus targets antigen receptor activation, particularly the T-cell receptor. Rapamycin exerts its immunosuppressive effect in a mechanism that is distinct from CsA. Early studies on rapamycin in *Saccharomyces cerevisiae* identified the 12 kDa immunophilin FKBP-12 as its intracellular receptor. It is this FKBP-12/drug complex that is the functional immunosuppressant *in vivo*. The mode of action of rapamycin is conserved from yeast to humans and numerous studies over the years have identified homologous targets for the FKBP-12/rapamycin complex in yeast and mammals.[102] The FKBP-12/rapamycin complex binds to and inhibits the kinase TOR, which is an abbreviation of Target of Rapamycin. This molecule is conserved through evolution and operates to regulate molecules downstream of PI3K, as described above.

Two important derivatives of rapamycin have been generated — called CCI-779 (developed by Wyeth) and RAD001 (SDZ RAD, everolimus, Certican; developed by Novartis).[100] Both agents have shown anti-proliferative activity in a variety of cancer cell lines. These derivatives are in clinical trials as cancer therapeutics. While both agents have shown toxicities, these are to be expected. Preliminary results suggest that CCI-779 is able to induce some response and prolong progression-free survival from breast and renal-cell carcinoma. RAD001 as well as being useful as an immunosuppressant could also inhibit the growth of Epstein-Barr virus immortalised B-cells in a mouse model. These two agents are likely to be clinically useful and demonstrate the efficacy of targeting PI3K regulated pathways.

7.5. Drug combinations

It is important to remember that PI3K inhibitors are unlikely to be the much talked about 'magic bullet' for cancer. The example of rapamycin indicates that inhibitors of PI3K downstream molecules are likely to be extremely useful for manipulating the immune system as well as for treating cancer. PI3K activity has also been implicated in a range of other diseases such as diabetes and respiratory disease, so specific inhibitors could have a wide range of applications. As mentioned above, the development of inhibitors of specific PI3K isoforms demonstrates that PI3K is both a valid and extremely important drug target.

Despite this optimism, it seems likely that PI3K inhibitors will be most effective when used in combination with other drugs. Cancer chemotherapy, to date, shows that combinations of drugs are often more effective than a single drug since targeting the tumour via multiple mechanisms avoids 'escape mutants' and can maintain patients in remission considerably longer. One study has shown that PI3K inhibitors can work well with the cytotoxic drugs paclitaxel.[144] Other work shows that even drugs such as rapamycin and LY294002, can cooperate to inhibit cell proliferation.[115] This suggests that targeting multiple molecules within a signalling network allows the reduction of the concentration of drugs required for phenotypic changes or potential therapeutic benefit. Thus, combinations with conventional drugs, or cocktails of signalling inhibitors may well be a fruitful area for success with PI3K therapeutics.

8. Concluding Comments

It is clear that PI3K plays a role in regulating cell survival, cell size, energy provision and cell proliferation. Various aspects of these phenotypes have been conserved throughout evolution and demonstrated in model systems. The critical role for PI3K depends on cell type, the subunit of PI3K under study, and perhaps even animal species. Consequently, it is possible to build up the following picture. Cells vary in their apoptotic threshold and some cell types appear more prone to death than others. In such cases, inhibiting PI3K may cause cell death but other cells require additional signals, such as high levels of c-myc expression. In these cells, activation of these parallel survival pathways (e.g. c-myc) allows other important roles of PI3K to be elucidated, since cells that cannot survive cannot do anything. This appears to be the case in studies where a low amount of PI3K activity is required, such as genetic models in the fly. In these examples, regulation of cell size which may be related to energy provision are elucidated.

Cell proliferation requires all of these other processes to be intact. A cell that is dead cannot proliferate. A cell that cannot maintain and even increase its size cannot divide. A cell without the necessary energy provision also loses its capacity to increase in number. There is also evidence that PI3K can regulate the expression and steady state levels of various important cell cycle molecules. Thus, the ultimate requirement for PI3K in cell proliferation may result from a number of effects of PI3K, including inhibition of cell death, increased energy provision, increased cell size, and activation of key cell cycle regulators. Only when all these PI3K-dependent signalling pathways are in place is cell division possible. As cancer often results in the alteration of many or all of these processes, specific molecules that target PI3K or its downstream pathways are valuable. These are already in the clinic in the form of inhibitors of tyrosine kinases and inhibitors of TOR. Furthermore, the development of these agents is ongoing.

Acknowledgements

This chapter describes the work of many, many researchers. For this reason, it has not been possible to include all the references to original literature. I have quoted many reviews. My apologies to any scientist I have failed to mention. No offense is deliberate.

The Brennan lab is supported by the Leukaemia Research Fund, UK, by Tenovus, the Cancer Charity and by Leukaemia Research Appeal for Wales.

I would like to thank Elaine Breslin, for help with the manuscript, and Angharad Shore for critical reading of this chapter.

References

1. D.A. Fruman, L.E. Rameh and L.C. Cantley, Phosphoinositide binding domains: embracing 3-phosphate, *Cell* **97** (1999) 817–820.
2. S.J. Isakoff, T. Cardozo, J. Andreev, Z. Li, K.M. Ferguson, R. Abagyan, M.A. Lemmon, A. Aronheim and E.Y. Skolnik, Identification and analysis

of PH domain-containing targets of phosphatidylinositol 3-kinase using a novel in vivo assay in yeast, *EMBO J.* **17** (1998) 5374–5387.
3. B. Vanhaesebroeck and D.R. Alessi, The PI3K-PDK1 connection: more than just a road to PKB, *Biochem. J.* **346** (Pt 3) (2000) 561–576.
4. D.A. Fruman, R.E. Meyers and L.C. Cantley, Phosphoinositide kinases, *Annu. Rev. Biochem.* **67** (1998) 481–507.
5. S.G. Ward and D.A. Cantrell, Phosphoinositide 3-kinases in T lymphocyte activation, *Curr. Opin. Immunol.* **13** (2001) 332–338.
6. B.H. Nelson and D.M. Willerford, Biology of the interleukin-2 receptor, *Adv. Immunol.* **70** (1998) 1–81.
7. P.J. Coffer, J. Jin and J.R. Woodgett, Protein kinase B (c-Akt): a multifunctional mediator of phosphatidylinositol 3-kinase activation, *Biochem. J.* **335** (Pt 1) (1998) 1–13.
8. E. Astoul, S. Watton and D. Cantrell, The dynamics of protein kinase B regulation during B cell antigen receptor engagement, *J. Cell Biol.* **145** (1999) 1511–1520.
9. P.S. Costello, M. Gallagher and D.A. Cantrell, Sustained and dynamic inositol lipid metabolism inside and outside the immunological synapse, *Nat. Immunol.* **3** (2002) 1082–1089.
10. J. Harriague and G. Bismuth, Imaging antigen-induced PI3K activation in T cells, *Nat. Immunol.* **3** (2002) 1090–1096.
11. A. Mora, D. Komander, D.M. van Aalten and D.R. Alessi, PDK1, the master regulator of AGC kinase signal transduction, *Semin. Cell Dev. Biol.* **15** (2004) 161–170.
12. P.J. Parker and S.J. Parkinson, AGC protein kinase phosphorylation and protein kinase C, *Biochem. Soc. Trans.* **29** (2001) 860–863.
13. J.R. Bayascas and D.R. Alessi, Regulation of Akt/PKB Ser473 phosphorylation, *Mol. Cell.* **18** (2005) 143–145.
14. D.D. Sarbassov, D.A. Guertin, S.M. Ali and D.M. Sabatini, Phosphorylation and regulation of Akt/PKB by the rictor-mTOR complex, *Science* **307** (2005) 1098–1101.
15. D. Stokoe, L.R. Stephens, T. Copeland, P.R. Gaffney, C.B. Reese, G.F. Painter, A.B. Holmes, F. McCormick and P.T. Hawkins, Dual role of phosphatidylinositol-3,4,5-trisphosphate in the activation of protein kinase B, *Science* **277** (1997) 567–570.
16. D.R. Alessi, S.R. James, C.P. Downes, A.B. Holmes, P.R. Gaffney, C.B. Reese and P. Cohen, Characterization of a 3-phosphoinositide-dependent protein kinase which phosphorylates and activates protein kinase Balpha, *Curr. Biol.* **7** (1997) 261–269.

17. B.J. Collins, M. Deak, J.S. Arthur, L.J. Armit and D.R. Alessi, In vivo role of the PIF-binding docking site of PDK1 defined by knock-in mutation, *EMBO J.* **22** (2003) 4202–4211.
18. M.R. Williams, J.S. Arthur, A. Balendran, J. van der Kaay, V. Poli, P. Cohen and D.R. Alessi, The role of 3-phosphoinositide-dependent protein kinase 1 in activating AGC kinases defined in embryonic stem cells, *Curr. Biol.* **10** (2000) 439–448.
19. D.A. Fruman, Phosphoinositide 3-kinase and its targets in B-cell and T-cell signaling, *Curr. Opin. Immunol.* **16** (2004) 314–320.
20. C.I. Smith, T.C. Islam, P.T. Mattsson, A.J. Mohamed, B.F. Nore and M. Vihinen, The Tec family of cytoplasmic tyrosine kinases: mammalian Btk, Bmx, Itk, Tec, Txk and homologs in other species, *Bioessays* **23** (2001) 436–446.
21. P.L. Schwartzberg, L.D. Finkelstein and J.A. Readinger, TEC-family kinases: regulators of T-helper-cell differentiation, *Nat. Rev. Immunol.* **5** (2005) 284–295.
22. H. Suzuki, Y. Terauchi, M. Fujiwara, S. Aizawa, Y. Yazaki, T. Kadowaki and S. Koyasu, Xid-like immunodeficiency in mice with disruption of the p85alpha subunit of phosphoinositide 3-kinase, *Science* **283** (1999) 390–392.
23. J.D. Thomas, P. Sideras, C.I. Smith, I. Vorechovsky, V. Chapman and W.E. Paul, Colocalization of X-linked agammaglobulinemia and X-linked immunodeficiency genes, *Science* **261** (1993) 355–358.
24. D.J. Rawlings, D.C. Saffran, S. Tsukada, D.A. Largaespada, J.C. Grimaldi, L. Cohen, R.N. Mohr, J.F. Bazan, M. Howard, N.G. Copeland, *et al.*, Mutation of unique region of Bruton's tyrosine kinase in immunodeficient XID mice, *Science* **261** (1993) 358–361.
25. M. Fukuda, T. Kojima, H. Kabayama and K. Mikoshiba, Mutation of the pleckstrin homology domain of Bruton's tyrosine kinase in immunodeficiency impaired inositol 1,3,4,5-tetrakisphosphate binding capacity, *J. Biol. Chem.* **271** (1996) 30303–30306.
26. M. Turner and D.D. Billadeau, VAV proteins as signal integrators for multi-subunit immune-recognition receptors, *Nat. Rev. Immunol.* **2** (2002) 476–486.
27. V.L. Tybulewicz, Vav-family proteins in T-cell signalling, *Curr. Opin. Immunol.* **17** (2005) 267–274.
28. M.J. Kim, E. Kim, S.H. Ryu and P.G. Suh, The mechanism of phospholipase C-gamma1 regulation, *Exp. Mol. Med.* **32** (2000) 101–109.
29. S.G. Rhee, Regulation of phosphoinositide-specific phospholipase C, *Annu. Rev. Biochem.* **70** (2001) 281–312.

30. E.M. Schaeffer and P.L. Schwartzberg, Tec family kinases in lymphocyte signaling and function, *Curr. Opin. Immunol.* **12** (2000) 282–288.
31. K.U. Birkenkamp and P.J. Coffer, Regulation of cell survival and proliferation by the FOXO (Forkhead box, class O) subfamily of Forkhead transcription factors, *Biochem. Soc. Trans.* **31** (2003) 292–297.
32. S.C. Kozma and G. Thomas, Regulation of cell size in growth, development and human disease: PI3K, PKB and S6K, *Bioessays* **24** (2002) 65–71.
33. C.M. Coelho and S.J. Leevers, Do growth and cell division rates determine cell size in multicellular organisms? *J. Cell Sci.* **113** (Pt 17) (2000) 2927–2934.
34. L.K. MacDougall, M.E. Gagou, S.J. Leevers, E. Hafen and M.D. Waterfield, Targeted expression of the class II phosphoinositide 3-kinase in Drosophila melanogaster reveals lipid kinase-dependent effects on patterning and interactions with receptor signaling pathways, *Mol. Cell Biol.* **24** (2004) 796–808.
35. P.M. Finan and M.J. Thomas, PI 3-kinase inhibition: a therapeutic target for respiratory disease, *Biochem. Soc. Trans.* **32** (2004) 378–382.
36. K. Okkenhaug and B. Vanhaesebroeck, PI3K in lymphocyte development, differentiation and activation, *Nat. Rev. Immunol.* **3** (2003) 317–330.
37. B. Vanhaesebroeck, K. Ali, A. Bilancio, B. Geering and L.C. Foukas, Signalling by PI3K isoforms: insights from gene-targeted mice, *Trends Biochem. Sci.* **30** (2005) 194–204.
38. D.A. Fruman, F. Mauvais-Jarvis, D.A. Pollard, C.M. Yballe, D. Brazil, R.T. Bronson, C.R. Kahn and L.C. Cantley, Hypoglycaemia, liver necrosis and perinatal death in mice lacking all isoforms of phosphoinositide 3-kinase p85 alpha, *Nat. Genet.* **26** (2000) 379–382.
39. Y. Terauchi, Y. Tsuji, S. Satoh, H. Minoura, K. Murakami, A. Okuno, K. Inukai, T. Asano, Y. Kaburagi, K. Ueki, H. Nakajima, T. Hanafusa, Y. Matsuzawa, H. Sekihara, Y. Yin, J.C. Barrett, H. Oda, T. Ishikawa, Y. Akanuma, I. Komuro, M. Suzuki, K. Yamamura, T. Kodama, H. Suzuki, S. Koyasu, S. Aizawa, K. Tobe, Y. Fukui, Y. Yazaki and T. Kadowaki, Increased insulin sensitivity and hypoglycaemia in mice lacking the p85 alpha subunit of phosphoinositide 3-kinase, *Nat. Genet.* **21** (1999) 230–235.
40. T. Fukao, M. Tanabe, Y. Terauchi, T. Ota, S. Matsuda, T. Asano, T. Kadowaki, T. Takeuchi and S. Koyasu, PI3K-mediated negative feedback regulation of IL-12 production in DCs, *Nat. Immunol.* **3** (2002) 875–881.
41. L. Bi, I. Okabe, D.J. Bernard and R.L. Nussbaum, Early embryonic lethality in mice deficient in the p110beta catalytic subunit of PI 3-kinase, *Mamm. Genome.* **13** (2002) 169–172.

42. L. Bi, I. Okabe, D.J. Bernard, A. Wynshaw-Boris and R.L. Nussbaum, Proliferative defect and embryonic lethality in mice homozygous for a deletion in the p110alpha subunit of phosphoinositide 3-kinase, *J. Biol. Chem.* **274** (1999) 10963–10968.
43. T. Shioi, P.M. Kang, P.S. Douglas, J. Hampe, C.M. Yballe, J. Lawitts, L.C. Cantley and S. Izumo, The conserved phosphoinositide 3-kinase pathway determines heart size in mice, *EMBO J.* **19** (2000) 2537–2548.
44. K. Okkenhaug, A. Bilancio, G. Farjot, H. Priddle, S. Sancho, E. Peskett, W. Pearce, S.E. Meek, A. Salpekar, M.D. Waterfield, A.J. Smith and B. Vanhaesebroeck, Impaired B and T cell antigen receptor signaling in p110delta PI 3-kinase mutant mice, *Science* **297** (2002) 1031–1034.
45. S.T. Jou, N. Carpino, Y. Takahashi, R. Piekorz, J.R. Chao, D. Wang and J.N. Ihle, Essential, nonredundant role for the phosphoinositide 3-kinase p110delta in signaling by the B-cell receptor complex, *Mol. Cell. Biol.* **22** (2002) 8580–8591.
46. E. Clayton, G. Bardi, S.E. Bell, D. Chantry, C.P. Downes, A. Gray, L.A. Humphries, D. Rawlings, H. Reynolds, E. Vigorito and M. Turner, A crucial role for the p110delta subunit of phosphatidylinositol 3-kinase in B cell development and activation, *J. Exp. Med.* **196** (2002) 753–763.
47. K. Ali, A. Bilancio, M. Thomas, W. Pearce, A.M. Gilfillan, C. Tkaczyk, N. Kuehn, A. Gray, J. Giddings, E. Peskett, R. Fox, I. Bruce, C. Walker, C. Sawyer, K. Okkenhaug, P. Finan and B. Vanhaesebroeck, Essential role for the p110delta phosphoinositide 3-kinase in the allergic response, *Nature* **431** (2004) 1007–1011.
48. T. Sasaki, J. Irie-Sasaki, R.G. Jones, A.J. Oliveira-dos-Santos, W.L. Stanford, B. Bolon, A. Wakeham, A. Itie, D. Bouchard, I. Kozieradzki, N. Joza, T.W. Mak, P.S. Ohashi, A. Suzuki and J.M. Penninger, Function of PI3Kgamma in thymocyte development, T cell activation, and neutrophil migration, *Science* **287** (2000) 1040–1046.
49. Z. Li, H. Jiang, W. Xie, Z. Zhang, A.V. Smrcka and D. Wu, Roles of PLC-beta2 and -beta3 and PI3Kgamma in chemoattractant-mediated signal transduction, *Science* **287** (2000) 1046–1049.
50. E. Hirsch, V.L. Katanaev, C. Garlanda, O. Azzolino, L. Pirola, L. Silengo, S. Sozzani, A. Mantovani, F. Altruda and M.P. Wymann, Central role for G protein-coupled phosphoinositide 3-kinase gamma in inflammation, *Science* **287** (2000) 1049–1053.
51. L.M. Webb, E. Vigorito, M.P. Wymann, E. Hirsch and M. Turner, Cutting edge: T cell development requires the combined activities of the

p110{gamma} and p110{delta} catalytic isoforms of phosphatidylinositol 3-kinase, *J. Immunol.* **175** (2005) 2783–2787.
52. M.A. Lawlor, A. Mora, P.R. Ashby, M.R. Williams, V. Murray-Tait, L. Malone, A.R. Prescott, J.M. Lucocq and D.R. Alessi, Essential role of PDK1 in regulating cell size and development in mice, *Embo. J.* **21** (2002) 3728–3738.
53. H.J. Hinton, D.R. Alessi and D.A. Cantrell, The serine kinase phosphoinositide-dependent kinase 1 (PDK1) regulates T cell development, *Nat. Immunol.* **5** (2004) 539–545.
54. A. Mora, C. Lipina, F. Tronche, C. Sutherland and D.R. Alessi, Deficiency of PDK1 in liver results in glucose intolerance, impairment of insulin-regulated gene expression and liver failure, *Biochem. J.* **385** (2005) 639–648.
55. Z.Z. Yang, O. Tschopp, A. Baudry, B. Dummler, D. Hynx and B.A. Hemmings, Physiological functions of protein kinase B/Akt, *Biochem. Soc. Trans.* **32** (2004) 350–354.
56. M.P. Scheid and J.R. Woodgett, PKB/AKT: functional insights from genetic models, *Nat. Rev. Mol. Cell. Biol.* **2** (2001) 760–768.
57. M.J. Parsons, R.G. Jones, M.S. Tsao, B. Odermatt, P.S. Ohashi and J.R. Woodgett, Expression of active protein kinase B in T cells perturbs both T and B cell homeostasis and promotes inflammation, *J. Immunol.* **167** (2001) 42–48.
58. R.G. Jones, M. Parsons, M. Bonnard, V.S. Chan, W.C. Yeh, J.R. Woodgett and P.S. Ohashi, Protein kinase B regulates T lymphocyte survival, nuclear factor kappaB activation, and Bcl-X(L) levels in vivo, *J. Exp. Med.* **191** (2000) 1721–1734.
59. S. Malstrom, E. Tili, D. Kappes, J.D. Ceci and P.N. Tsichlis, Tumor induction by an Lck-MyrAkt transgene is delayed by mechanisms controlling the size of the thymus, *Proc. Natl. Acad. Sci. USA* **98** (2001) 14967–14972.
60. T. Shioi, J.R. McMullen, P.M. Kang, P.S. Douglas, T. Obata, T.F. Franke, L.C. Cantley and S. Izumo, Akt/protein kinase B promotes organ growth in transgenic mice, *Mol. Cell. Biol.* **22** (2002) 2799–2809.
61. H. Cho, J.L. Thorvaldsen, Q. Chu, F. Feng and M.J. Birnbaum, Akt1/PKBalpha is required for normal growth but dispensable for maintenance of glucose homeostasis in mice, *J. Biol. Chem.* **276** (2001) 38349–38352.
62. R.S. Garofalo, S.J. Orena, K. Rafidi, A.J. Torchia, J.L. Stock, A.L. Hildebrandt, T. Coskran, S.C. Black, D.J. Brees, J.R. Wicks, J.D. McNeish and K.G. Coleman, Severe diabetes, age-dependent loss of adipose tissue,

and mild growth deficiency in mice lacking Akt2/PKB beta, *J. Clin. Invest.* **112** (2003) 197–208.

63. W.S. Chen, P.Z. Xu, K. Gottlob, M.L. Chen, K. Sokol, T. Shiyanova, I. Roninson, W. Weng, R. Suzuki, K. Tobe, T. Kadowaki and N. Hay, Growth retardation and increased apoptosis in mice with homozygous disruption of the Akt1 gene, *Genes Dev.* **15** (2001) 2203–2208.
64. Z.Z. Yang, O. Tschopp, M. Hemmings-Mieszczak, J. Feng, D. Brodbeck, E. Perentes and B.A. Hemmings, Protein kinase B alpha/Akt1 regulates placental development and fetal growth, *J. Biol. Chem.* **278** (2003) 32124–32131.
65. S.S. Bae, H. Cho, J. Mu and M.J. Birnbaum, Isoform-specific regulation of insulin-dependent glucose uptake by Akt/protein kinase B, *J. Biol. Chem.* **278** (2003) 49530–49536.
66. J. Chen, S. De, D.S. Damron, W.S. Chen, N. Hay and T.V. Byzova, Impaired platelet responses to thrombin and collagen in AKT-1-deficient mice, *Blood* **104** (2004) 1703–1710.
67. R.M. Easton, H. Cho, K. Roovers, D.W. Shineman, M. Mizrahi, M.S. Forman, V.M. Lee, M. Szabolcs, R. de Jong, T. Oltersdorf, T. Ludwig, A. Efstratiadis and M.J. Birnbaum, Role for Akt3/protein kinase Bgamma in attainment of normal brain size, *Mol. Cell. Biol.* **25** (2005) 1869–1878.
68. O. Tschopp, Z.Z. Yang, D. Brodbeck, B.A. Dummler, M. Hemmings-Mieszczak, T. Watanabe, T. Michaelis, J. Frahm and B.A. Hemmings, Essential role of protein kinase B(gamma) (PKB(gamma)/Akt3) in postnatal brain development but not in glucose homeostasis, *Development* **132** (2005) 2943–2954.
69. X.D. Peng, P.Z. Xu, M.L. Chen, A. Hahn-Windgassen, J. Skeen, J. Jacobs, D. Sundararajan, W.S. Chen, S.E. Crawford, K.G. Coleman and N. Hay, Dwarfism, impaired skin development, skeletal muscle atrophy, delayed bone development, and impeded adipogenesis in mice lacking Akt1 and Akt2, *Genes. Dev.* **17** (2003) 1352–1365.
70. P.S. Hammerman, C.J. Fox, M.J. Birnbaum and C.B. Thompson, Pim and Akt oncogenes are independent regulators of hematopoietic cell growth and survival, *Blood* **105** (2005) 4477–4483.
71. R. Dhand, K. Hara, I. Hiles, B. Bax, I. Gout, G. Panayotou, M.J. Fry, K. Yonezawa, M. Kasuga and M.D. Waterfield, PI 3-kinase: structural and functional analysis of intersubunit interactions, *EMBO J.* **13** (1994) 511–521.
72. B.L. Craddock, E.A. Orchiston, H.J. Hinton and M.J. Welham, Dissociation of apoptosis from proliferation, protein kinase B activation, and BAD

phosphorylation in interleukin-3-mediated phosphoinositide 3-kinase signaling, *J. Biol. Chem.* **274** (1999) 10633–10640.
73. Q.X. Zhang, I.D. Davis and G.S. Baldwin, Controlled overexpression of selected domains of the P85 subunit of phosphatidylinositol 3-kinase reverts v-Ha-Ras transformation, *Biochim. Biophys. Acta.* **1312** (1996) 207–214.
74. N. Rahimi, E. Tremblay and B. Elliott, Phosphatidylinositol 3-kinase activity is required for hepatocyte growth factor-induced mitogenic signals in epithelial cells, *J. Biol. Chem.* **271** (1996) 24850–24855.
75. S. Tachiiri, K. Sasai, N. Oya and M. Hiraoka, Enhanced cell killing by overexpression of dominant-negative phosphatidylinositol 3-kinase subunit, Deltap85, following genotoxic stresses, *Jpn. J. Cancer Res.* **91** (2000) 1314–1318.
76. A. Klippel, C. Reinhard, W.M. Kavanaugh, G. Apell, M.A. Escobedo and L.T. Williams, Membrane localization of phosphatidylinositol 3-kinase is sufficient to activate multiple signal-transducing kinase pathways, *Mol. Cell Biol.* **16** (1996) 4117–4127.
77. K. Reif, C.D. Nobes, G. Thomas, A. Hall and D.A. Cantrell, Phosphatidylinositol 3-kinase signals activate a selective subset of Rac/Rho-dependent effector pathways, *Curr. Biol.* **6** (1996) 1445–1455.
78. K. Reif, B.M. Burgering and D.A. Cantrell, Phosphatidylinositol 3-kinase links the interleukin-2 receptor to protein kinase B and p70 S6 kinase, *J. Biol. Chem.* **272** (1997) 14426–14433.
79. S. Ward, Y. Sotsios, J. Dowden, I. Bruce and P. Finan, Therapeutic potential of phosphoinositide 3-kinase inhibitors, *Chem. Biol.* **10** (2003) 207–213.
80. M. Ui, T. Okada, K. Hazeki and O. Hazeki, Wortmannin as a unique probe for an intracellular signalling protein, phosphoinositide 3-kinase, *Trends Biochem. Sci.* **20** (1995) 303–307.
81. C.J. Vlahos, W.F. Matter, K.Y. Hui and R.F. Brown, A specific inhibitor of phosphatidylinositol 3-kinase, 2-(4-morpholinyl)-8-phenyl-4H-1-benzopyran-4-one (LY294002), *J. Biol. Chem.* **269** (1994) 5241–5248.
82. S.P. Davies, H. Reddy, M. Caivano and P. Cohen, Specificity and mechanism of action of some commonly used protein kinase inhibitors, *Biochem. J.* **351** (2000) 95–105.
83. S. Roche, M. Koegl and S.A. Courtneidge, The phosphatidylinositol 3-kinase alpha is required for DNA synthesis induced by some, but not all, growth factors, *Proc. Natl. Acad. Sci. USA* **91** (1994) 9185–9189.
84. B. Vanhaesebroeck and M.D. Waterfield, Signaling by distinct classes of phosphoinositide 3-kinases, *Exp. Cell. Res.* **253** (1999) 239–254.

85. D.A. Cross, D.R. Alessi, P. Cohen, M. Andjelkovich and B.A. Hemmings, Inhibition of glycogen synthase kinase-3 by insulin mediated by protein kinase B, *Nature* **378** (1995) 785–789.
86. D. Cantrell, Protein kinase B (Akt) regulation and function in T lymphocytes, *Semin. Immunol.* **14** (2002) 19–26.
87. J.R. Woodgett, Recent advances in the protein kinase B signaling pathway, *Curr. Opin. Cell Biol.* **17** (2005) 150–157.
88. J. Downward, PI 3-kinase, Akt and cell survival, *Semin. Cell Dev. Biol.* **15** (2004) 177–182.
89. K.U. Birkenkamp and P.J. Coffer, FOXO transcription factors as regulators of immune homeostasis: molecules to die for? *J. Immunol.* **171** (2003) 1623–1629.
90. S. Ogg, S. Paradis, S. Gottlieb, G.I. Patterson, L. Lee, H.A. Tissenbaum and G. Ruvkun, The Fork head transcription factor DAF-16 transduces insulin-like metabolic and longevity signals in C. elegans, *Nature* **389** (1997) 994–999.
91. L.P. Kane, V.S. Shapiro, D. Stokoe and A. Weiss, Induction of NF-kappaB by the Akt/PKB kinase, *Curr. Biol.* **9** (1999) 601–604.
92. O.N. Ozes, L.D. Mayo, J.A. Gustin, S.R. Pfeffer, L.M. Pfeffer and D.B. Donner, NF-kappaB activation by tumour necrosis factor requires the Akt serine-threonine kinase, *Nature* **401** (1999) 82–85.
93. J.A. Romashkova and S.S. Makarov, NF-kappaB is a target of AKT in anti-apoptotic PDGF signalling, *Nature* **401** (1999) 86–90.
94. X. Li, J.C. Tupper, D.D. Bannerman, R.K. Winn, C.J. Rhodes and J.M. Harlan, Phosphoinositide 3 kinase mediates Toll-like receptor 4-induced activation of NF-kappaB in endothelial cells, *Infect Immun.* **71** (2003) 4414–4420.
95. N. Sizemore, S. Leung and G.R. Stark, Activation of phosphatidylinositol 3-kinase in response to interleukin-1 leads to phosphorylation and activation of the NF-kappaB p65/RelA subunit, *Mol. Cell. Biol.* **19** (1999) 4798–4805.
96. K. Du and M. Montminy, CREB is a regulatory target for the protein kinase Akt/PKB, *J. Biol. Chem.* **273** (1998) 32377–32379.
97. J. Jin, N. Anthopoulos, B. Wetsch, R.C. Binari, D.D. Isaac, D.J. Andrew, J.R. Woodgett and A.S. Manoukian, Regulation of Drosophila tracheal system development by protein kinase B, *Dev. Cell* **1** (2001) 817–827.
98. J. Downward and S.J. Leevers, Trachealess — a new transcription factor target for PKB/Akt, *Dev. Cell* **1** (2001) 726–728.
99. J.I. Bardos and M. Ashcroft, Hypoxia-inducible factor-1 and oncogenic signalling, *Bioessays* **26** (2004) 262–269.

100. M.A. Bjornsti and P.J. Houghton, The TOR pathway: a target for cancer therapy, *Nat. Rev. Cancer* **4** (2004) 335–348.
101. T. Powers and P. Walter, Regulation of ribosome biogenesis by the rapamycin-sensitive TOR-signaling pathway in *Saccharomyces cerevisiae*, *Mol. Biol. Cell.* **10** (1999) 987–1000.
102. R.T. Abraham and G.J. Wiederrecht, Immunopharmacology of rapamycin, *Annu. Rev. Immunol.* **14** (1996) 483–510.
103. N. Hay and N. Sonenberg, Upstream and downstream of mTOR, *Genes Dev.* **18** (2004) 1926–1945.
104. K. Hara, Y. Maruki, X. Long, K. Yoshino, N. Oshiro, S. Hidayat, C. Tokunaga, J. Avruch and K. Yonezawa, Raptor, a binding partner of target of rapamycin (TOR), mediates TOR action, *Cell* **110** (2002) 177–189.
105. P. Brennan, J.W. Babbage, G. Thomas and D. Cantrell, p70(s6k) integrates phosphatidylinositol 3-kinase and rapamycin-regulated signals for E2F regulation in T lymphocytes, *Mol. Cell. Biol.* **19** (1999) 4729–4738.
106. S. Volarevic, M.J. Stewart, B. Ledermann, F. Zilberman, L. Terracciano, E. Montini, M. Grompe, S.C. Kozma and G. Thomas, Proliferation, but not growth, blocked by conditional deletion of 40S ribosomal protein S6, *Science* **288** (2000) 2045–2047.
107. E.C. Holland, N. Sonenberg, P.P. Pandolfi and G. Thomas, Signaling control of mRNA translation in cancer pathogenesis, *Oncogene* **23** (2004) 3138–3144.
108. A.C. Gingras, B. Raught and N. Sonenberg, mTOR signaling to translation, *Curr. Top. Microbiol. Immunol.* **279** (2004) 169–197.
109. A.C. Gingras, B. Raught and N. Sonenberg, eIF4 initiation factors: effectors of mRNA recruitment to ribosomes and regulators of translation, *Annu. Rev. Biochem.* **68** (1999) 913–963.
110. A.C. Gingras, S.G. Kennedy, M.A. O'Leary, N. Sonenberg and N. Hay, 4E-BP1, a repressor of mRNA translation, is phosphorylated and inactivated by the Akt(PKB) signaling pathway, *Genes. Dev.* **12** (1998) 502–513.
111. A. Kauffmann-Zeh, P. Rodriguez-Viciana, E. Ulrich, C. Gilbert, P. Coffer, J. Downward and G. Evan, Suppression of c-Myc-induced apoptosis by Ras signalling through PI(3)K and PKB, *Nature* **385** (1997) 544–548.
112. P. Brennan, A.M. Mehl, M. Jones and M. Rowe, Phosphatidylinositol 3-kinase is essential for the proliferation of lymphoblastoid cells, *Oncogene* **21** (2002) 1263–1271.
113. P. Brennan, J.W. Babbage, B.M. Burgering, B. Groner, K. Reif and D.A. Cantrell, Phosphatidylinositol 3-kinase couples the interleukin-2 receptor to the cell cycle regulator E2F, *Immunity* **7** (1997) 679–689.

114. L. Banerji, J. Glassford, N.C. Lea, N.S. Thomas, G.G. Klaus and E.W. Lam, BCR signals target p27(Kip1) and cyclin D2 via the PI3K signalling pathway to mediate cell cycle arrest and apoptosis of WEHI 231 B cells, *Oncogene* **20** (2001) 7352–7367.
115. E.M. Breslin, P.C. White, A.M. Shore, M. Clement and P. Brennan, LY294002 and rapamycin co-operate to inhibit T-cell proliferation, *Br. J. Pharmacol.* **144** (2005) 791–800.
116. B.C. Fox, T.E. Crew and M.J. Welham, Phosphoinositide 3-kinases can act independently of p27Kip1 to regulate optimal IL-3-dependent cell cycle progression and proliferation, *Cell. Signal.* **17** (2005) 473–487.
117. A. Martino, J.H.t. Holmes, J.D. Lord, J.J. Moon and B.H. Nelson, Stat5 and Sp1 regulate transcription of the cyclin D2 gene in response to IL-2, *J. Immunol.* **166** (2001) 1723–1729.
118. P.C. White, A.M. Shore, M. Clement, J. McLaren, I. Soeiro, E.W. Lam and P. Brennan, Regulation of cyclin D2 and the cyclin D2 promoter by protein kinase A and CREB in lymphocytes. *Oncogene* (2005) **doi**: 10.1038/sj.onc.1209255.
119. J.J. Moon and B.H. Nelson, Phosphatidylinositol 3-kinase potentiates, but does not trigger, T cell proliferation mediated by the IL-2 receptor, *J. Immunol.* **167** (2001) 2714–2723.
120. M. Stahl, P.F. Dijkers, G.J. Kops, S.M. Lens, P.J. Coffer, B.M. Burgering and R.H. Medema, The forkhead transcription factor FoxO regulates transcription of p27Kip1 and Bim in response to IL-2, *J. Immunol.* **168** (2002) 5024–5031.
121. P.F. Dijkers, R.H. Medema, C. Pals, L. Banerji, N.S. Thomas, E.W. Lam, B.M. Burgering, J.A. Raaijmakers, J.W. Lammers, L. Koenderman and P.J. Coffer, Forkhead transcription factor FKHR-L1 modulates cytokine-dependent transcriptional regulation of p27(KIP1), *Mol. Cell. Biol.* **20** (2000) 9138–9148.
122. J. Slingerland and M. Pagano, Regulation of the cdk inhibitor p27 and its deregulation in cancer, *J. Cell. Physiol.* **183** (2000) 10–17.
123. J. Nourse, E. Firpo, W.M. Flanagan, S. Coats, K. Polyak, M.H. Lee, J. Massague, G.R. Crabtree and J.M. Roberts, Interleukin-2-mediated elimination of the p27Kip1 cyclin-dependent kinase inhibitor prevented by rapamycin, *Nature* **372** (1994) 570–573.
124. K.A. Frauwirth and C.B. Thompson, Regulation of T lymphocyte metabolism, *J. Immunol.* **172** (2004) 4661–4665.
125. K.A. Frauwirth, J.L. Riley, M.H. Harris, R.V. Parry, J.C. Rathmell, D.R. Plas, R.L. Elstrom, C.H. June and C.B. Thompson, The CD28 signaling pathway regulates glucose metabolism, *Immunity* **16** (2002) 769–777.

126. J. Luo, B.D. Manning and L.C. Cantley, Targeting the PI3K-Akt pathway in human cancer: rationale and promise, *Cancer Cell* **4** (2003) 257–262.
127. I. Vivanco and C.L. Sawyers, The phosphatidylinositol 3-Kinase AKT pathway in human cancer, *Nat. Rev. Cancer* **2** (2002) 489–501.
128. P.A. Futreal, L. Coin, M. Marshall, T. Down, T. Hubbard, R. Wooster, N. Rahman and M.R. Stratton, A census of human cancer genes, *Nat. Rev. Cancer* **4** (2004) 177–183.
129. R. Parsons, Human cancer, PTEN and the PI-3 kinase pathway, *Semin. Cell Dev. Biol.* **15** (2004) 171–176.
130. S. Cooray, The pivotal role of phosphatidylinositol 3-kinase-Akt signal transduction in virus survival, *J. Gen. Virol.* **85** (2004) 1065–1076.
131. P. Brennan, Signalling events regulating lymphoid growth and survival, *Semin. Cancer Biol.* **11** (2001) 415–421.
132. Y. Liu, Y. Wang, M. Yamakuchi, S. Masuda, T. Tokioka, S. Yamaoka, I. Maruyama and I. Kitajima, Phosphoinositide 3-kinase-PKB/Akt pathway activation is involved in fibroblast Rat-1 transformation by human T-cell leukemia virus type I tax, *Oncogene* **20** (2001) 2514–2526.
133. A. Polychronis, H.D. Sinnett, D. Hadjiminas, H. Singhal, J.L. Mansi, D. Shivapatham, S. Shousha, J. Jiang, D. Peston, N. Barrett, D. Vigushin, K. Morrison, E. Beresford, S. Ali, M.J. Slade and R.C. Coombes, Preoperative gefitimib versus gefitinib and anastrozole in postmenopausal patients with oestrogen-receptor positive and epidermal-growth-factor receptor-positive primary breast cancer: a double-blind placebo-controlled phase II randomised trial, *Lancet Oncol.* **6** (2005) 838–891.
134. C. Sadhu, B. Masinovsky, K. Dick, C.G. Sowell and D.E. Staunton, Essential role of phosphoinositide 3-kinase delta in neutrophil directional movement, *J. Immunol.* **170** (2003) 2647–2654.
135. Y.L. Chen, P.Y. Law and H.H. Loh, Inhibition of akt/protein kinase B signaling by naltrindole in small cell lung cancer cells, *Cancer Res.* **64** (2004) 8723–8730.
136. R. Bortul, P.L. Tazzari, A.M. Billi, G. Tabellini, I. Mantovani, A. Cappellini, T. Grafone, G. Martinelli, R. Conte and A.M. Martelli, Deguelin, A PI3K/AKT inhibitor, enhances chemosensitivity of leukaemia cells with an active PI3K/AKT pathway, *Br. J. Haematol.* **129** (2005) 677–686.
137. D. Komander, G.S. Kular, A.W. Schuttelkopf, M. Deak, K.R. Prakash, J. Bain, M. Elliott, M. Garrido-Franco, A.P. Kozikowski, D.R. Alessi and D.M. van Aalten, Interactions of LY333531 and other bisindolyl maleimide inhibitors with PDK1, *Structure* (Camb.) **12** (2004) 215–226.

138. D. Komander, G.S. Kular, J. Bain, M. Elliott, D.R. Alessi and D.M. Van Aalten, Structural basis for UCN-01 (7-hydroxystaurosporine) specificity and PDK1 (3-phosphoinositide-dependent protein kinase-1) inhibition, *Biochem. J.* **375** (2003) 255–262.
139. L. Yang, H.C. Dan, M. Sun, Q. Liu, X.M. Sun, R.I. Feldman, A.D. Hamilton, M. Polokoff, S.V. Nicosia, M. Herlyn, S.M. Sebti and J.Q. Cheng, Akt/protein kinase B signaling inhibitor-2, a selective small molecule inhibitor of Akt signaling with antitumor activity in cancer cells overexpressing Akt, *Cancer Res.* **64** (2004) 4394–4399.
140. S.N. Sehgal, H. Baker and C. Vezina, Rapamycin (AY-22,989), a new antifungal antibiotic. II. Fermentation, isolation and characterization, *J. Antibiot. (Tokyo)* **28** (1975) 727–732.
141. C. Vezina, A. Kudelski and S.N. Sehgal, Rapamycin (AY-22,989), a new antifungal antibiotic. I. Taxonomy of the producing streptomycete and isolation of the active principle, *J. Antibiot. (Tokyo)* **28** (1975) 721–726.
142. C.J. Watson, P.J. Friend, N.V. Jamieson, T.W. Frick, G. Alexander, A.E. Gimson and R. Calne, Sirolimus: a potent new immunosuppressant for liver transplantation, *Transplantation* **67** (1999) 505–509.
143. I.D. van Mourik and D.A. Kelly, Immunosuppressive drugs in paediatric liver transplantation, *Paediatr. Drugs* **3** (2001) 43–60.
144. L. Hu, J. Hofmann, Y. Lu, G.B. Mills and R.B. Jaffe, Inhibition of phosphatidylinositol 3′-kinase increases efficacy of paclitaxel in vitro and in vivo ovarian cancer models, *Cancer Res.* **62** (2002) 1087–1092.

5 FOXO transcription factors

Silvia Fernandez de Mattos, Jan J. Brosens and Eric W.-F. Lam

1. Introduction/Summary

FOXO proteins have emerged as key regulators of cell fate decisions in response to hormonal, growth factor and environmental cues. This reputation of FOXO proteins is based on their ability to control diverse and opposing cellular responses such as cell cycle progression, DNA repair, defense against oxidative damage and apoptosis. Not surprisingly, the transcriptional activity of FOXO proteins is carefully controlled at multiple levels, although most prominently by signal-induced, post-translational modifications. The selection of gene sets activated by nuclear FOXO is regulated by targeted acetylation and deacetylation as well as by the interaction with other transcription factors and co-activators. Activation of the PI3K/PKB pathway by growth factors results in phosphorylation of nuclear FOXO at specific sites, which is further modulated by additional phosphorylation events mediated by other kinases. PKB-dependent phosporylation will disable FOXO proteins by inactivating the FOXO nuclear translocation signal and by exporting these transcription factors from the nucleus in a CRM1- and 14-3-3-dependent process. Interaction with the ubiquitin ligase Skp2 in the cytoplasm will earmark PKB-phosphorylated FOXO for proteasomal degradation. Conversely, stress signals and phosphorylation of cytoplasmic FOXO by JNK can reverse the nuclear-cytoplasmic shuttle, resulting in nuclear import and transcriptional activation of FOXO proteins. This chapter reviews the mechanisms that fine-tunes FOXO transcriptional activity as well as the evidence that links disruption of these control systems to human disease.

2. Forkhead Transcription Factors

The winged helix of Forkhead box (FOX) class of transcription factors constitute a family of structurally related transcriptional activators (approximately 100 members) that have been identified in species ranging from yeast to human. The first member of this transcription factor class was identified in 1989 as a nuclear homeotic gene (*fork head* gene or *fkh*) involved in *Drosophila melanogaster* embryonic development.[1] In 1990, Weigel and Jackle identified the forkhead box DNA-binding domain.[2] The discovery of this forkhead motif led to the definition of a new family of transcription factors, the FOX class of transcription factors (*F*orkhead b*OX*-containing protein).

Members of the Forkhead superfamily encode a subgroup of the helix-turn-helix class of proteins, and are characterised by the presence of a conserved 110-aa DNA-binding domain, called the 'forkhead' domain or 'winged helix' domain. The arrangement of loops (or wings) connecting the β-strands that flank one of the three α-helices gives rise to a butterfly-like appearance, hence their alternative name of 'winged-helix' transcription factors (Fig. 1). This is a relatively invariant structure, with most of the amino acids being conserved between family members. Other portions of the Fox proteins, which encode for trans-activation or trans-repression domains, are highly divergent between the different family members.

A phylogenetic analysis has resulted in the definition of 15 classes for all known Fox proteins, so these transcription factors are classified in terms of structure not function. The Fox factors play essential roles in a wide range of cellular processes, and are involved in the development of tissues from all three germ layers.[3] Among other processes, Foxs are important in:

- the establishment of the body axis (Foxb1b, Foxa2),
- specification of the left-right axis (Foxh1, Foxj1),
- ciliogenesis (Foxj1),

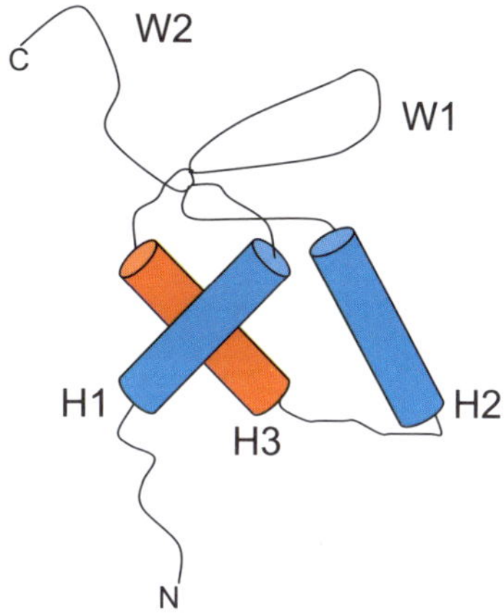

Fig. 1 The structure shows the helix-turn-helix motif viewed from the side facing away from the DNA with two helices, H1 and H2 (blue), stacked on top of the recognition helix H3 (red). The recognition helix H3 lies in the major groove of DNA. Also shown are the two loops or 'wings' (W1 and W2).

- speech and language development (FOXP2),
- inner ear development and hearing (Foxi1),
- cranio-pharyngeal development (FoxE1), and
- hepatic and pancreatic metabolism and foregut development (Foxa).

2.1. Nomenclature of forkhead transcription factors

Fox (Forkhead box) has been adopted as the unified symbol for all chordate winged helix/forkhead transcription factors. Fox proteins have been assigned to 17 individual subclasses according to the amino acid sequence of their conserved forkhead domain.[4] Subclasses have been designated by a letter (class A to Q), and within each subclass proteins have an Arabic numeral. Therefore, the actual name of any Fox protein is 'Fox, subclass N, member X' (e.g. Foxd3). Abbreviations for the chordate Fox proteins will contain all uppercase letters for human (e.g. FOXD3), only the first letter capitalised for mouse (e.g. Foxd3), and the first and subclass letters capitalised for all other chordates (e.g. FoxD3). Lowercase letters are used to distinguish between virtually identical proteins (e.g. Foxa4a and

Foxa4b), presumably derived from duplicated genes, a case commonly found in polyploid species like *Xenopus laevis*. An updated phylogenetic tree can be downloaded from the Forkhead gene nomenclature website (http://www.biology.pomona.edu/fox.html).

3. FoxO Transcription Factors

FoxO (forkhead box O; forkhead members of the O class) are transcription factors that function under the control of growth factors and insulin signalling. Initial investigations leading to the identification of Forkhead transcription factors as key effectors of PI3K-PKB signalling began in the nematode worm *Caenorhabditis elegans* (*C. elegans*). This organism feeds on bacteria and usually lives for a few days. However, when environmental conditions are adverse, worms can enter a so-called Dauer or 'non-ageing' larval stage where they can survive for months. Dauer represents a complex phenotype, characterised by life span extension, metabolic changes, increased stress resistance and delayed sexual reproduction. This is regulated by a neuroendocrine pathway which has been extensively studied by genetic analysis.

FoxO transcription factors are the mammalian homologues of the *C. elegans* DAF-16 transcription factor. In this organism, DAF-16 has been shown to act as a key nuclear target of signalling by the insulin-like receptor DAF-2. Mutants in *daf-2* and *age-1* (a homologue of the catalytic subunit of mammalian PI3K) arrest development at the dauer stage. Null mutations in *daf-16* suppress the effects of mutations in *daf-2* or *age-1*, and lack of *daf-16* bypasses the need for this insulin receptor-like signalling pathway.[5] The principal role of DAF-2/AGE-1 signalling is thus to antagonise DAF-16, a Forkhead transcription factor.

Within the large family of Fox factors, three human orthologues of DAF-16 have been identified: FKHR (Forkhead in rhabdomyosarcoma), FKHR-L1 (FKHR-like 1) and AFX (acute-lymphocytic-leukaemia-1 fused gene from chromosome X). Interestingly, these proteins were initially identified at chromosomal break points in several human tumours.[6,7] With the newly proposed nomenclature for Fox factors, these proteins have recently been renamed: FKHR corresponds to FoxO1, FKHR-L1 to FoxO3a and AFX to FoxO4, respectively.

The FoxO factors all function as transcriptional activators and interact as monomers with the core consensus DNA sequence TTGTTTAC,[8] through binding of helix-3 to the major groove of DNA. The residues flanking this core contribute to the DNA binding specificity.

3.1. FOXO6

A novel forkhead domain gene has been cloned and characterised that is highly related to the FoxO group of transcription factors, and has therefore been designated FoxO6.[9] Embryonic expression of FoxO6 is most apparent in the developing brain, and FoxO6 is expressed in a specific temporal and spatial pattern. Therefore, it is probably involved in regulation of specific cellular differentiation.

Interestingly, FoxO6 clearly differs from FoxO1 and FoxO3 in its shuttling properties. In transfection studies, FoxO6 coupled to GFP shows an unexpectedly high nuclear localisation after stimulation with growth factors, in contrast to the predominant cytosolic localisation of FoxO1 and FoxO3. Structure function analysis of FoxO6 compared with its group members shows that the overall homology is high, but surprisingly a highly conserved region containing multiple phosphorylation sites is lacking. Mutation analysis revealed that the dramatically reduced nucleo-cytoplasmic shuttling of FoxO6 is due to the absence of this functional domain.[9]

4. Signalling to FOXO Proteins

In their function as key controllers of development and differentiation, Fox proteins might be predicted to be nuclear targets for signal transduction. Surprisingly, this has remained a poorly investigated field, with the probable exception of members of the FoxO class. FOXO factors are regulated by several signal transduction cascades. The main regulator of FOXO function is the PI3K/PKB pathway, whereas FOXO function is 'fine-tuned' by the protein kinases CK1 (formerly known as casein kinase 1) and the DYRK1A (dual-specificity regulated kinase 1A) pathway. These and other kinases regulate the intracellular localisation of FOXO proteins by phosphorylating FOXO factors within different intramolecular domains. We will discuss in detail the different upstream kinases involved in regulation of FOXO activity.

4.1. Phosphorylation of FOXO by the PI3K/PKB pathway

As discussed, mechanistic evidence suggesting that FOXO proteins are regulated by PKB was initially obtained from genetic studies in *C. elegans*. Analysis of the DAF-16 amino acid sequence revealed four sites that correspond to the consensus PKB phosphorylation site (RXRXXS/T, where 'X' denotes any residue), suggesting that DAF-16 and other Fox transcription factors might be direct targets of PKB.

It was not until 1999 that two different research groups demonstrated that FOXO function can be mediated by direct PKB phosphorylation in mammalian systems.[10,11] Subsequently, FOXOs have been found to be phosphorylated *in vivo* on multiple threonine and serine residues (T1 and T2, and S1 to S5, Fig. 2). Three of these phosphorylated residues, one N-terminal threonine (T1) and two C-terminal serines (S1 and S2) lie within a consensus motif for phosphorylation by PKB. Indeed, FOXOs have been shown to be phosphorylated by PKB in these sites both *in vitro* and *in vivo*.[10–13] The *C. elegans* FoxO, DAF-16, contains an additional overlapping PKB motif

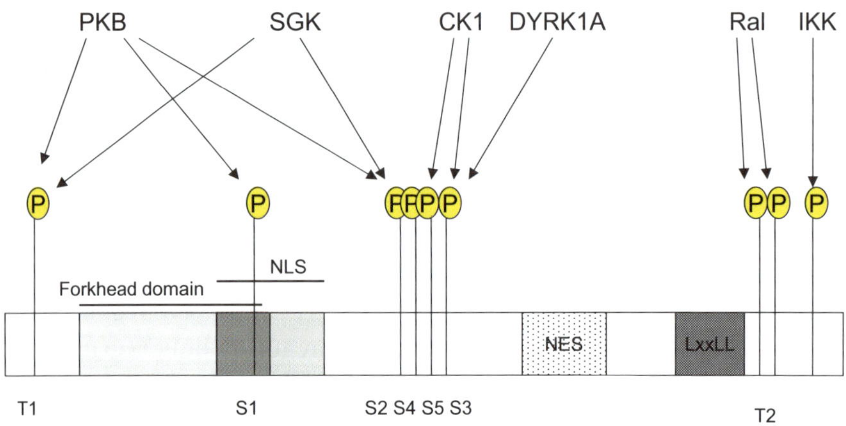

Fig. 2 Structure and phosphorylation sites of the FoxO transcription factors. NLS: Nuclear localisation sequence, NES: nuclear export sequence.

in the forkhead domain. Interestingly, FOXO6 lacks the C-terminal PKB consensus motif.

All FOXO proteins have been shown to require the consensus T1 and S1 PKB sites to translocate from the nucleus to the cytosol.[14] These two phosphorylated residues are essential for translocation, as they influence the NLS (nuclear localisation sequence) function,[14] and the asociation with 14-3-3 proteins.[10] The N-terminal PKB site (T1) also regulates the association of FOXO with the coactivator p300.[15] In the absence of growth factors, p300 binds to the N-terminal part of FOXO3a. Growth factor addition and phosphorylation of the T1 motif disrupts the interaction with p300, thereby preventing the acetylation of FOXO3a and directly influencing FOXO-mediated transactivation.

Phosphorylation of the C-terminal PKB site (S2) accelerates nuclear export of the FOXO factors.[16] The domain surrounding this PKB site is important for the efficiency of translocation and is a target for additional signalling events, as will be discussed below. The lack of a third PKB site at the proper position in the protein has clear functional consequences, as highlighted by FOXO6, which lacks this site.[9] Under most conditions, FOXO6 is located in the nucleus, whereas FOXO1 and FOXO3 have a predominant cytosolic localisation after PKB activation.

4.2. Phosphorylation of FOXO by SGK

The serum- and glucocorticoid inducible kinases (SGKs) form a family of serine/threonine kinases that are activated in response to a variety of extracellular stimuli. SGK is closely related to PKB and its activation is also dependent on PI3K activity, although it is not recruited to the plasma membrane by PIP_2 and PIP_3, since it does not contain a pleckstrin homology domain.

It has been shown that SGK1 promotes cell survival and that it does so in part by phosphorylating and inactivating FOXO3a.[17] However, SGK and PKB display differences with respect to the efficacy with which they phosphorylate the three regulatory sites on FOXO3a. While both kinases can phosphorylate T1, SGK displays a marked preference for S2 whereas PKB favours S1. These findings suggest that SGK and PKB may coordinately regulate the function of FOXOs by phosphorylating these transcription factors at distinct sites.

4.3. Phosphorylation of FOXO by CK1

CK1 is a serine/threonine kinase, of which at least seven isoforms have been identified. All CK1 proteins contain a highly homologous kinase domain, and their kinase activity is negatively regulated by autophosphorylation. Importantly, CK1 recognises and phosphorylates motifs that have been 'primed', i.e. motifs that already contain a phosphorylated serine or threonine residue. Phosphorylation of S2 by PKB forms a consensus sequence for phosphorylation by CK1, allowing it to phosphorylate S4, which in turn primes the CK1-catalysed phosphorylation of S5.[16]

4.4. Phosphorylation of FOXO by DYRK1A

DYRK1A belongs to the dual-specificity tyrosine-phosphorylated and regulated kinase (DYRK1A) family of kinases. DYRKs are serine/threonine kinases that have tyrosine autophosphorylation activity. A DYRK1A site (S3), adjacent to the second CK1 site, completes the stretch of serine residues that form an acidic patch upon phosphorylation. Phosphorylation of S3 by DYRK1A decreases the ability of FOXO1 to stimulate gene transactivation and reduces the proportion of FOXO1 present in the nucleus.[18] This residue is highly phosphorylated in unstimulated cells and its phosphorylation is not altered by stimulation with insulin-like growth factor-1 (IGF-1), by treatment with inhibitors of PI3K, or by mutants of the PKB and CK1 sites.[16] This indicates that regulation of the DYRK1A site is involved in the subcellular localisation of FOXO proteins, despite the fact that the site is independent of PKB and CKI activity.

4.5. Phosphorylation of FOXO by the Ras-Ral pathway

A pathway involving Ras signalling to the Ral GTPase has been implicated in the phosphorylation of FOXO4 at T2 (Thr-447/Thr-451).[19] Phosphorylation of these residues does not affect PKB-induced nuclear exit but is required for normal transcriptional activity. However, since these residues are in a region that is poorly conserved between FOXO isoforms, it is not clear whether regulation by the Ras/Ral pathway is restricted to FOXO4.

4.6. Phosphorylation of FOXO by IKK

Signalling through the IκB kinase (IKK) and NFκB has been established as a key survival and anti-apoptotic pathway. The primary form of the transcription factor NFκB is a heterodimer of p50 and p65 that is retained as a latent form in the cytoplasm of resting cells by IκB, an inhibitor of NFκB. NFκB is activated by stimulation of an IKK complex wich phosphorylates IκB and triggers its ubiquitination-dependent degradation through the proteasome pathway. This leads to nuclear translocation of activated NFκB and upregulation of its target genes.

It has recently been shown that IKKβ can also regulate the nuclear-cytoplasmic localisation of FOXO transcription factors.[20] This observation came first as an inverse correlation between the levels of nuclear FOXO3A and IKKβ in tumour samples. It was consequently shown that IKKβ interacts with and phosphorylates FOXO3a, independently of PKB. IKKβ phosphorylation of FOXO3a represses FOXO3a transcriptional activity and promotes FOXO3a nuclear exclusion and proteolysis through a ubiquitin-dependent proteasome pathway, which leads to cell cycle progression and proliferation.

IKK is capable of phosphorylating the C-terminal region of FOXO3a and Ser-644 is the major target site. Inactivation of FOXO3a through ubiquitination and degradation is dependent on Ser-644 phosphorylation. Interestingly, over-expression of FOXO3a can override IKKβ stimulation and suppress tumourigenesis in mice, indicating that in this setting FOXO3a is acting as a tumour suppressor.

5. Cellular Relocation Mechanism

The phosphorylation of FOXOs by PKB, SGK, CK1, DYRK1A and IKK is crucial for their localisation within the cell. The PKB site located in the forkhead domain acts as a trigger, and the other sites function as 'fine-tuners' of the shuttling behaviour. The stretch of four phosphorylated serine residues is thought to form an acidic patch, facilitating nuclear export.[16] Insertion of the four serine residues from FoxO3 into the corresponding region of FoxO6 rescues the ability of FoxO6 to shuttle efficiently from the nucleus to the cytosol, emphasising the importance of this region in FoxO relocation.[9]

The first clues to the mechanism of phosphorylation-dependent relocation of FOXOs came from the observation that phosphorylation of FOXOs by PKB generates two consensus binding sites for 14-3-3 proteins.[21] 14-3-3 chaperone proteins form homo- and hetero-dimeric U-shaped structures that bind to a large number of phosphorylated signalling molecules through the large central channel in the centre of the U. 14-3-3 proteins regulate interactions between the bound proteins and other molecules and also influence the intracellular localisation of bound ligands. When 14-3-3 proteins complex to nuclear ligands (such as FoxO), the bound proteins relocalise and become sequestered in the cytoplasm by several mechanims.[22] 14-3-3 proteins interact with FoxO, and this interaction is dependent on phosphorylation of the two PKB consensus sites.[10] The importance of the 14-3-3 interaction has been demonstrated *in vitro* with DAF-16.[23] DAF-16 is efficiently phosphorylated by PKB *in vitro*, which promotes DAF-16 association with 14-3-3. This phosphorylation-dependent interaction with 14-3-3 is sufficient to displace DAF-16 from DNA. Thus, binding of 14-3-3 proteins to FOXO that has been phosphorylated by PKB is required for the release of FOXO factors from DNA and their subsequent translocation to the cytoplasm.

Although 14-3-3 binding appears to be sufficient to displace FOXO factors from the DNA, the localisation of FOXO proteins within the nucleus is regulated by the presence of nuclear localisation signal (NLS) and nuclear export signal (NES) motifs in these transcription factors. The presence of an NLS is a prerequisite for maintaining proteins in the nucleus, whereas an NES maintains proteins in the cytosol.

The sequences required for efficient nuclear localisation of FOXO proteins have been mapped to a series of basic amino acids residues that constitute a non-classical NLS located C-terminal to the DNA-binding domain.[14,24] A conserved PKB site is conserved within this motif, and a FOXO4 mutant in which this site was converted to a glutamic acid to mimic phosphorylation exhibited reduced nuclear localisation.[14] This suggests that PKB phosphorylation may promote export in part by disrupting NLS function. Additionally, subsequent binding by 14-3-3 may also act to mask NLS function by blocking the interaction of FOXO with the nuclear import machinery.

FOXO transcription factors also contain leucine-rich regions that can function as NES.[14,24] Mutation of NES motifs in FOXO3a blocks FOXO export out of the nuclear compartment, even when sites of PKB phosphorylation and 14-3-3 interaction are intact.[22] However, an intact NES is not sufficient for export, because a FOXO3a mutant that can no longer bind 14-3-3 remains localised to the nuclear compartment, although this mutant retains intact NES sequences. These results suggest that effective nuclear export requires both 14-3-3 binding and an intact NES on FOXO.

The transport of FoxO proteins through the nuclear pore is dependent on active-transport mechanisms that require adaptor proteins to mediate either import or export. These adaptors are called importin or exportin receptors, respectively. Importins and exportins recognise specific NLSs and NESs present in the transported protein. FoxO proteins have both an NLS and an NES. Phosphorylation by kinases and interactions with other proteins modulate the effectiveness of these NLSs and NESs, which form the base of FoxO shuttling.

As mentioned, a PKB phosphorylation motif is present within the FoxONLS. Phosphorylation of the PKB site introduces a negative charge, which might influence the NLS. Moreover, the phosphorylated residue might sterically hinder NLS function. Once FoxO proteins become phosphorylated they interact with the small GTPase Ran[16] and the exportin CRM1,[14] and they are transported through the nuclear pore complex towards the cytosol.

In summary, translocation of FoxOs from the nucleus to the cytosol can be divided into several sequential steps:

- Phosphorylation of the PKB site in the forkhead domain leads to disruption of the NLS function and disruption of DNA binding. Phosphorylation leads to accessibility of the N- and C-terminal PKB sites.
- The N- and C-terminal PKB sites are phosphorylated. Phosphorylation of the C-terminal PKB site by SGK or PKB leads to the phosphorylation of the stretch of three serine residues.

- The current phosphorylation state induces high-affinity binding to 14-3-3 proteins.
- The phosphorylated FoxO protein in a complex with Ran and CRM1 is transported through the nuclear pore complex towards the cytoplasm.

Disruption of one of the above steps results in disruption of nuclear export, and accumulation of FoxO proteins in the nucleus.

6. FOXO Protein Processing

In addition to the shuttling system explained before, cytosolic phosphorylated FoxO proteins are degraded by the ubiquitin-proteasome system, providing the cell with a double negative regulation of FoxO factors. It has been described that activated PKB results in a decrease in FOXO1 and FOXO3a protein levels.[25] This reduction in FOXO1 and FOXO3a protein levels can be blocked by treatment with proteasome inhibitors, suggesting that the proteasome system has a role in regulating FOXO activity. In fact, FOXO1 is ubiquitinated and degraded by the proteasome in response to insulin, a process that is PKB-dependent.[26]

Besides processing by the ubiquitin system, FOXOs are also processed by proteases. FOXO3a is a substrate for caspase-3-like proteases,[27] and the sequence identified as the cleavage site appears to be conserved in FOXO1 and FOXO4. Interestingly, androgens can initiate protease-mediated cleavage of FOXO1 into two fragments.[28] The C-terminal fragment contains the DNA-binding domain and a large portion of the transactivation domain, and acts as a dominant negative. Via this protease-mediated mechanism, androgens are believed to deregulate FOXO1-mediated processes.

7. Transcriptional Activity of FoxO Proteins

The activity of FoxO proteins can be regulated by re-positioning the protein in the cell as discussed previously. Also, direct regulation of transcriptional capacity by protein modification has been reported. Finally, FoxO proteins interact and cooperate with other proteins, and these interactions can cause changes in FoxO-mediated transactivation, as well as influence other transcription factors (Fig. 3).

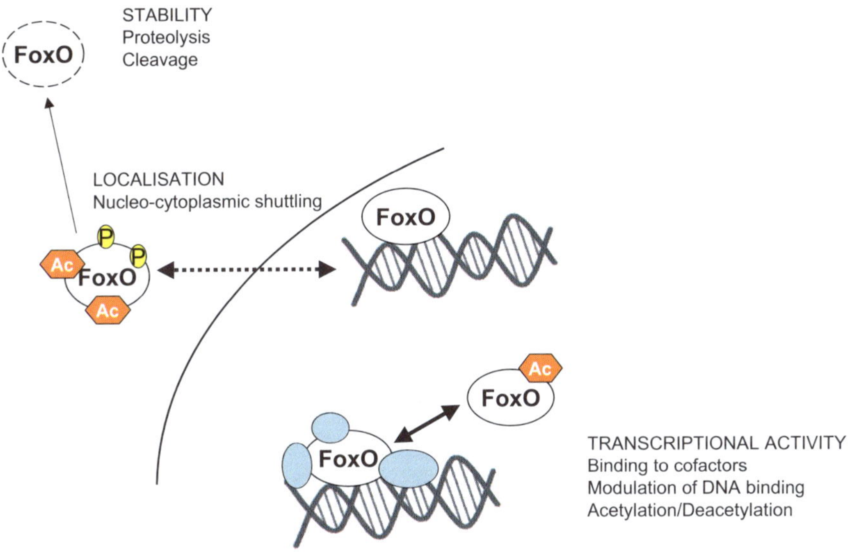

Fig. 3 Levels of FoxO regulation.

7.1. Interactions with nuclear hormone receptors

Besides regulating transcription via the direct interaction between FoxOs and DNA, FoxO factors have a DNA-binding-independent effect on transcription through their interaction with nuclear receptors. The nuclear receptor superfamily includes receptors for steroid hormones, vitamins, fatty acids, and other lipophilic signalling molecules. Nuclear receptors mostly function as ligand-dependent transcription factors. FoxO proteins can influence nuclear receptor transactivation by activating or repressing transcription (depending on the nuclear receptor involved), while conversely interaction of nuclear receptors with FoxO factors can inhibit FoxO-mediated transcription.[29,30]

Nuclear receptors interacting with FoxO factors include the oestrogen receptor (ER),[30] the progesterone receptor (PR),[29] the androgen receptor (AR),[31] the thyroid hormone receptor,[29] the glucocorticoid receptor,[29] the retinoic acid receptor (RAR),[29] the peroxisome-proliferator-activated receptor (PPARγ),[32] and hepatocyte nuclear factor-4 (HNF-4).[33]

Steroid receptors are retained in the cytoplasm, and upon ligand binding they are released from their chaperones, dimerise, enter the nucleus and bind to specific response elements in the regulatory region of target genes. In contrast with steroid receptors, non-steroid receptors are mainly located in the nucleus. In the absence of ligand, they are associated with histone-deacetylase-containing complexes tethered through co-repressors. This process results in chromatin compactation and silencing of the target genes. Upon ligand binding, the co-repressor-binding interface is destabilised, which leads to their dissociation. Subsequently, nuclear receptors bound to their responsive elements facilitate recruitment of the transcriptional machinery.

All FoxO factors contain an LxxLL domain located in the C-terminal region of the protein, and it has been suggested that this region is responsible for interaction with nuclear receptors.[34] The mechanism underlying FoxO regulation of nuclear receptors and *vice versa* is very complex, and is influenced by multiple pathways. First, the interaction between nuclear receptors and FoxOs can be dependent or independent of ligand binding to nuclear receptors.[29] As an example, binding of FOXO1 to RAR is ligand-independent, whereas binding of FOXO1 to ER is enhanced by oestrogen. Secondly, FoxO phosphorylation can influence the interaction with nuclear receptors, as has been shown for FOXO1 and HNF-4.[33] Since phosphorylation of FoxO factors influences their subcellular localisation, it should be assumed that phosphorylation affects the interactions with steroid and non-steroid receptors. However, repression of FOXO1 activity by AR is independent of FOXO1 phosphorylation,[31] suggesting that the result on transactivation depends on the particular nuclear receptor involved.

Interactions between nuclear receptors and FOXO1, FOXO3a and FOXO4 have been documented, but to date nothing is known about the novel FoxO factor FoxO6, which also possesses the LxxLL motif. Since its localisation is mainly nuclear, it would provide nuclear receptors with a co-factor that is always present.

7.2. FoxO acetylation

The observation that in *C. elegans* the Sir2 deacetylase positively regulates life span in a DAF-16-dependent manner,[35] raised the interesting possibility that FoxO may be regulated by the mammalian sir orthologs, or Sirtuins

(Sirt1-7). Nuclear Sirtuins can deacetylate transcription factors, such as p53, while cytoplasmic Sirtuins deacetylate enzymes and structural proteins like tubulin. Two independent studies now expand the repertoire of Sirt1 targets by including FoxO.[36,37] These studies show that FoxO is acetylated in response to cellular stress by the acetyltransferase activity of the nuclear hormone receptor co-activators CBP and p300, while expression of Sirt1 causes FoxO deacetylation.

The functional consequences of this deacetylation process are less clear. While Motta and colleagues claim that deacetylation of FoxO factors is a general mechanism that deactivates FoxO transciptional activity,[37] Brunet and colleagues suggest that the effect of deacetylation is target gene specific, and thus regulates the balance between pro-apoptotic and cell cycle arresting genes.[36] What is generally agreed is that the net effect of deacetylation is to promote cellular survival under stress conditions, and to protect against apoptosis by fine-tuning FoxO-mediated transcription.

8. Functions of FOXO *In Vivo*

FoxO transcription factors have been associated with cell-type-specific effects on cell cycle, cell death, metabolism, DNA repair and protection against oxidative stress. Since most of their downstream targets have been identified *in vitro*, the *in vivo* extrapolation is problematic. However, data from studies in *C. elegans*, *Drosophila* and mice provide us with information regarding the *in vivo* function of FoxO factors.

8.1. FoxO in *C. elegans*

As discussed previously, DAF-16 is the FoxO homologue expressed in *C. elegans*, and is regulated by the DAF-2/AGE-1 signalling pathway. Life span extension, stress resistance and arrest at the dauer diapause stage are accomplished by direct activation of DAF-16.[5] Starvation, heat and oxidative stress all activate DAF-16, whereas nutrient-rich conditions deactivate DAF-16. In summary, DAF-16 regulates the allocation of resources in *C. elegans* as a response to a changing environment.[38]

8.2. FoxO in *Drosophila melanogaster*

The Drosophila FoxO homologue was identified recently and named dFOXO.[39–41] Ectopic expression of dFOXO results in a marked reduction in body size, caused by a decrease in cell number and cell size. This phenotype resembles that observed in starving larvae, and correlates well with the fact that nutrients and the insulin-PI3K pathway negatively regulate dFOXO.

dFoxO knockout flies are viable and of normal size, but show increased vulnerability to oxidative stress, suggesting that dFoxO may not be required for proper growth but provides protection against oxidative stress.[41] It has been suggested that under stressful conditions, such as nutrient deprivation, dFoxO is activated and regulates growth via the translation inhibitor d4E-BP.[39,41]

8.3. FoxO in *Mus musculus*

Each one of the FoxO genes has been disrupted in mice,[42] and the results show that the physiological roles of the different family members in mammals are quite diverse. FoxO1 homozygous null embryos die before birth due to incomplete vascular development, suggesting a crucial role for this transcription factor as a regulator of embryonic vessel formation. Both FoxO3a and FoxO4-null mice are viable. FoxO3a knockouts show haematological abnormalities, age-dependent infertility and abnormal ovarian follicular development due to premature follicular activation. FoxO3a is thus suggested to function at early stages of follicular growth as a suppressor of follicular activation. In contrast, histological analyses of FoxO4-null mice could not identify any consistent abnormalities.

These studies reveal an essential role for FoxO1 in embryonic vascular development and for FoxO3a in ovarian follicular physiology, whereas FoxO4 is seemingly non-essential. Much still remains to be determined regarding the role of FoxO family members at the cellular and organismal levels. The diverse functions of FoxO members may be due in part to differences in their transcriptional targets. The targets that mediate the effects of FoxO activity on cell survival, cell cycle control, DNA repair and protection against oxidative stress are just beginning to be identified.

9. FOXO Targets and Function

9.1. Regulation of cell cycle

Cell division is the process by which a cell duplicates its DNA content and cell mass to generate two daughter cells. In mammals, this response is essential for embryo development and for repair and renewal of tissues in the adult organism. The signalling pathways regulating FoxO activity have a clear role in regulating cell proliferation, suggesting the possible involvement of FoxO family members in the control of cell cycle progression.

Studies in a variety of cell types have confirmed that over-expression of a constitutively active form of FoxO induces arrest in the G1 phase of the cell cycle.[19,43] FoxO activity mediates G1 cell cycle arrest at least by inducing expression of the cell cycle inhibitor $p27^{Kip1}$.[44] $p27^{Kip1}$ likely represents a direct target of FoxO activity because multiple FoxO binding sites are present in the $p27^{Kip1}$ promoter. It has also been reported that FoxO-induced accumulation of $p27^{Kip1}$ involves stabilisation of the protein.[43] $p27^{Kip1}$ appears to be a major mediator of FoxO-induced G1 arrest because $p27^{Kip1}$-deficient fibroblasts are largely resistant to FoxO4-induced G1 arrest.

FOXO4 and FOXO3a also directly control transcription of the retinoblastoma-like p130 protein and cause upregulation of p130 protein expression.[45] Following FOXO-induced cell cycle arrest, cells enter G0 and become quiescent. This is shown by a change in phosphorylation of p130 to G0-specific forms and an increase in p130/E2F-4 complex formation. The suppression of the D-type cyclins has also been implicated in FOXO-mediated growth arrest.[46] Forced expression or conditional activation of FoxO factors leads to reduced protein expression of the D-type cyclins D1 and D2. Down-regulation of D-type cyclins involves a transcriptional repression mechanism and does not require $p27^{Kip1}$ function. Ectopic expression of cyclin D1 can partially overcome FoxO factor-induced cell cycle arrest, demonstrating that down-regulation of D-type cyclins represents a physiologically relevant mechanism of FoxO-induced cell cycle inhibition.

Contrary to the $p27^{Kip1}$ promoter, the promoter regions of the D-type cyclins do not contain any obvious FoxO-binding sites, suggesting that transcriptional regulation either takes place through the interaction with other transcription factors or is indirectly mediated through FoxO-dependent induction of a transcriptional repressor protein. In a recent report, this was shown to be the case for cyclin D2: FoxO3a activation will in turn induce the expression of the transcriptional repressor BCL6, that will bind to the cyclin D2 promoter culminating in the down-regulation of cyclin D2 transcription.[47]

FoxO proteins are also engaged as mediators of p21^{Cip1} activation by TGFβ,[48] which makes them a point of convergence for the PI3K and TGFβ pathways. FoxO1, FoxO3a and FoxO4 can bind specifically and directly to Smad3 and Smad4 and target the region of the p21^{Cip1} promoter that contains a consensus FoxO binding element next to a Smad binding element, resulting in the transactivation of p21^{Cip1}.[48] These FoxO-Smad complexes can be inhibited by FoxG1, a distinct Forkhead family member that is essential for neuroepithelial development. FoxG1 can recruit corepressors,[49] so its presence may turn a FoxO-Smad complex into a repressor complex.

Another cell cycle target of FoxO factors is cyclin G2,[50] an unconventional cyclin highly expressed in terminally differentiated cells. Unlike classical cyclins that promote cell cycle progression, cyclin G2 blocks cell cycle entry when over-expressed.[51] The *cyclin G2* promoter contains clustered FoxO consensus motifs that bind and are transactivated by FoxO factors. This suggests an additional mechanism for control of quiescence by FoxO proteins, that is the regulation of expression of cyclin G2.

There is evidence for a biphasic, complementary role of FoxO transcription factors during progression through the cell cycle. In G1, the PI3K pathway is active, inactivates FoxO, and due to the effect on the molecular targets discussed above this allows entry into S phase. On the contrary, during G2 phase PI3K is inactive, and FoxO transcription factors are again activated, which is necessary for cell cycle termination.[52] Interference with FoxO transcriptional activity induces accumulation of cells in G2/M, defective cytokinesis and delayed transition from M into G1.

The mechanism by which FoxO factors regulate cytokinesis and M/G1 transition include regulation of cyclin B and polo-like kinase (Plk) transcription[52] by direct binding to their promoter regions. This suggests that FoxO family members play a key role in controlling mitosis by regulating the expression of a set of genes that act as crucial mediators of M phase progression.

9.2. Regulation of apoptosis

Programmed cell death or apoptosis is an important mechanism for the elimination of cells harbouring mutations that will otherwise result in uncontrolled cell proliferation. The first activity to be adscribed to FoxO factors was the induction of apoptosis. The PI3K/PKB pathway acts in a variety of cell types to promote survival.[53,54] The observation that PKB phosphorylation inhibits FoxO transcriptional activity suggested that FoxO factors might play a role in apoptosis. Indeed, over-expression of a constitutive form of FoxO in which all three PKB sites have been mutated to alanine promotes apoptosis in a number of cell types.[10,55,56] Importantly, the pro-apoptotic activity of this mutant depends on DNA binding, suggesting that the effect on cell death is due to FoxO transcriptional activity.[13]

Subsequent analysis has shown that FoxO-induced apoptosis is mediated through transcriptional regulation of a number of target genes. One of the first targets to be identified was the death cytokine Fas ligand (FasL), a mediator of FoxO-induced cell death.[10] The FasL promoter contains putative FoxO binding elements, that have been shown to bind FoxO3a.[52] Once induced by FoxO, secreted FasL activates the cell surface death receptor Fas in a paracrine or autocrine manner, with subsequent caspase activation and resultant apoptosis.

In some cases, FoxO-induced cell death is independent of Fas signalling. The up-regulation of the pro-apoptotic Bcl-2 family member Bim plays a role in induction of apoptosis by FoxO upon cytokine withdrawal in haematopoietic cell lines.[56,57] Bim functions at the mitochondria, where it promotes release of cytochrome C, initiating a cascade of proteolytic events that are hallmarks of programmed cell death. Bim appears to be a direct target of FoxO, as FoxO-induced up-regulation of Bim mRNA is independent of new protein synthesis.[57] Interestingly, Bim-deficient haematopoietic stem cells are largely resistant to cell death induced by growth factor withdrawal or inhibition of the PI3K pathway.[56]

In breast cancer, FoxO3a has also been shown to up-regulate Bim.[58] In paclitaxel-sensitive breast cancer cells, paclitaxel treatment upregulates FoxO3a, which subsequently increases Bim expression. This will ultimately decrease cell survival and contribute to the tumour response to paclitaxel. In chronic myeloid leukaemia cells, FoxO3a has been implicated as a downstream effector of the cell cycle arrest induced by the BCR-ABL chemical inhibitor STI571.[59] Recently, FoxO3a has been shown to directly mediate induction of Bim when BCR-ABL activity is inhibited by STI571.[60]

FoxO transcription factors can also indirectly down-regulate the expression of genes involved in cell survival. FoxO4 has been shown to suppress the expression of Bcl-xL, a pro-survival member of the Bcl-2 family. This effect is mediated through the transcriptional repressor Bcl6.[61] As was the case for cyclin D2 regulation by FoxO3a, FoxO4 transactivates the *Bcl6* promoter, leading to an increase in Bcl6 expression, that in turn binds to the *Bcl-xL* promoter and represses its expression. Thus, by inducing cell death genes and indirectly supressing pro-survival genes, FoxO activation contributes to the initiation of programmed cell death.

9.3. Regulation of stress response and DNA repair

The first evidence that FoxOs might regulate the cellular response to environmental stress derives from studies in *C. elegans*, in which the FoxO ortholog DAF-16 was found to be induced by stress stimuli.[38] FoxOs have also been demonstrated to play a role in the stress response of mammalian cells, thereby providing protection from oxidative stress. Different studies have identified numerous FoxO-regulated stress-responsive genes involved in a variety of stress responses ranging from antioxidant defense to DNA repair. In particular, FoxO3a appears to enhance expression of the antioxidant enzymes mitochondrial manganese superoxide dismutase (MnSOD) and catalase.[62,63] Cells deficient in MnSOD were no longer protected from oxidant damage by ectopic expression of FOXO, suggesting that this is a critical defense pathway. In agreement with this, expression of FoxO3a has been shown to result in an increase in both hydrogen peroxide scavenging and oxidative stress resistance.[63]

One mechanism by which cells protect themselves against oxidative stress is by repairing the damage to their DNA and proteins that occurs upon exposure to environmental stress. Under conditions of DNA damage, cycling cells arrest at one of two critical cell cycle checkpoints (G1-S or G2-M phase of the cell cycle, before and after DNA replication) allowing the cells time to repair their DNA. The ability of FoxO family members to induce cell cycle arrest at the G1-S and G2-M checkpoints suggests a possible involvement of FoxO factors in DNA damage repair.

By gene array analysis, FOXO3a was found to modulate the expression of several genes that regulate the cellular response to stress at the G2-M checkpoint. The growth arrest and DNA damage response gene *Gadd45a* appeared to be a direct target of FOXO3a that mediates part of FOXO3a's effects on DNA repair.[64] The function of Gadd45 is to prevent the interaction of cdc2 with cyclin B.[65] FoxO3a and FoxO4 activate the *Gadd45* promoter through interaction with FoxO binding motifs, and FoxO3a-induced DNA repair was found to be partially impaired in Gadd45a-null fibroblasts. It has been suggested that under low stress conditions, FOXO3a may promote DNA repair, whereas under higher levels of stress, FOXO3a may induce a programme of cell death. This graded response to stress stimuli would protect cells from damage but also facilitate the removal of heavily damaged cells, resulting in an overall increase in the life span of the organism.

10. FOXO Proteins in Transformation and Diseases

Given the involvement of FoxO transcription factors in cellular responses such as cell cycle control, DNA repair and apoptosis, it is not surprising that these proteins may play a role in carcinogenesis. The signalling pathways that modulate FoxO activity have been clearly implicated in tumourigenesis. Ras, PI3K and PKB all function as oncogenes, and PTEN, the lipid phosphatase that opposes PI3K signalling in the cell, is a tumour suppressor frequently deleted in prostate cancer. Thus, FoxO family members function within a pathway whose dysregulation can lead to tumourigenesis.

In fact, the members of the FoxO subfamily were initially isolated as the sites of chromosomal break points present in human tumours such as rhabdomyosarcomas and acute leukaemias.[6,7] Chromosomal rearrangements that occurred during the oncogenic process result in the fusion of the FoxO transactivation domain to the DNA-binding domains of other unrelated transcription factors. In rhabdomyosarcomas, the transactivation domain of FoxO1 is fused to the DNA-binding domain of the transcription factors PAX3 or PAX7;[6] similarly, in acute leukaemias the transactivation domains of FoxO3 and FoxO4 are fused to the DNA-binding domain of the MLL factor.[7] These fusion products may result in a gain-of-function, dysregulating cell physiology due to excessive transcription of PAX and MLL target genes. Alternatively, the chromosomal translocation would lead to the loss of an allele of the FoxO gene, while the chimeric FoxO protein can interfere with the function of the remaining intact allele. This may result in a nearly complete loss of FoxO function, which may pre-dispose cells to dysregulated cell cycle control and impaired apoptosis that result in neoplasia.

Foxo1 has been shown to control two important processes in the pathogenesis of type 2 diabetes in mice: hepatic glucose production and cell compensation of insulin resistance. Foxo1 is thus a possible therapeutic target to alleviate insulin resistance and improve cell compensation. Treatments that lower Foxo1 levels should help alleviate the metabolic abnormalities associated with insulin resistance.[66]

So far, no diseases have been directly linked to mutations in specific FoxO transcription factors, perhaps owing to their more generalised function in the regulation of cellular proliferation and survival. FoxOs themselves could be considered as potential tumour suppressors, with an inhibitory mutation contributing to transformation. Future studies might identify disease-causing mutations in members of the FoxO subfamily, similar to those found for other Fox transcription factors.

11. Concluding Remarks

All FoxO members are regulated by multiple protein kinases, thereby serving as a transcriptional endpoint of several signalling cascades. Central to the regulation of FoxO factors is a shuttling system, which confines FoxO factors to either the nucleus or the cytosol. When members of the FoxO family of transcription factors are located in the nucleus, they activate expression of genes that can prevent proliferation and promote apoptosis. When cells are stimulated with growth factors, signalling pathways become activated that lead to phosphorylation of FoxO proteins. This causes their localisation to the cytoplasm and subsequent cell proliferation. In the absence of growth or survival signals, however, FoxO factors are retained in the nucleus resulting in the inhibition of cell division, as well as tumour suppression. Besides the regulation of intracellular localisation, growth factors influence FoxO association with the general transcription machinery and DNA binding itself, providing the cell with an additional pathway in the regulation of transcription.

In vitro and *in vivo* studies have identified an array of FoxO downstream targets (Fig. 4). These targets implicate FoxO in processes such as cell cycle regulation, cell death, protection from oxidative stress and metabolism. Inactivation of FoxO proteins perturbs the critical balance between cell proliferation and cell death and contributes to tumourigenesis by promoting cell growth and cell survival (Fig. 5).

Cell cycle	Apoptosis	Stress response and DNA repair
$p27^{Kip1}$	FasL	
p130	Bim	MnSOD
Cyclin D1	Bcl-XL (Bcl6)	Gadd45
Cyclin D2		
$p21^{Cip1}$		
Cyclin G2		
Cyclin B		
Plk		

Fig. 4 FoxO targets.

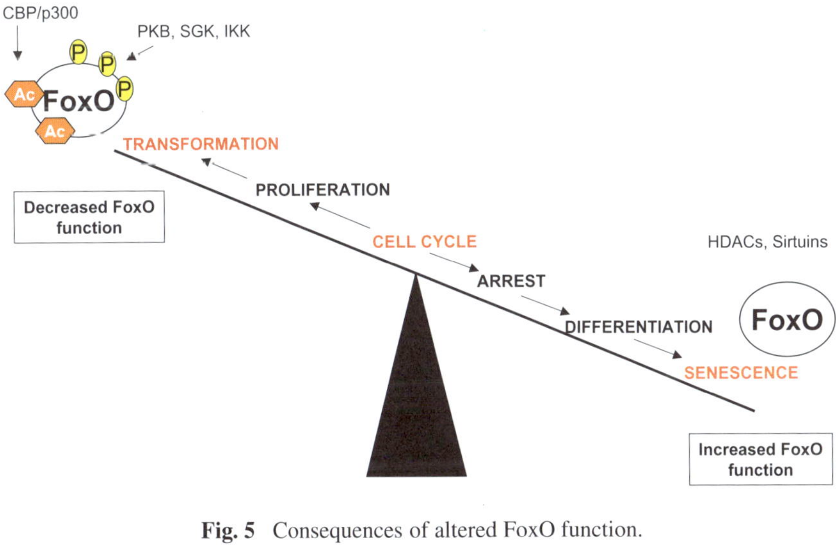

Fig. 5 Consequences of altered FoxO function.

References

1. D. Weigel, G. Jurgens, F. Kuttner, E. Seifert and H. Jackle, The homeotic gene fork head encodes a nuclear protein and is expressed in the terminal regions of the Drosophila embryo, *Cell* **57** (1989) 645–658.
2. D. Weigel and H. Jackle, The fork head domain: a novel DNA binding motif of eukaryotic transcription factors? *Cell* **63** (1990) 455–456.
3. O.J. Lehmann, J.C. Sowden, P. Carlsson, T. Jordan and S.S. Bhattacharya, Fox's in development and disease, *Trends Genet.* **19** (2003) 339–344.
4. K.H. Kaestner, W. Knochel and D.E. Martinez, Unified nomenclature for the winged helix/forkhead transcription factors, *Genes Dev.* **14** (2000) 142–146.
5. S. Ogg, S. Paradis, S. Gottlieb, G.I. Patterson, L. Lee, H.A. Tissenbaum and G. Ruvkun, The Fork head transcription factor DAF-16 transduces insulin-like metabolic and longevity signals in *C. elegans*, *Nature* **389** (1997) 994–999.
6. N. Galili, R.J. Davis, W.J. Fredericks, S. Mukhopadhyay, F.J. Rauscher, III, B.S. Emanuel, G. Rovera and F.G. Barr, Fusion of a fork head domain gene to PAX3 in the solid tumour alveolar rhabdomyosarcoma, *Nat. Genet.* **5** (1993) 230–235.
7. A. Borkhardt, R. Repp, O.A. Haas, T. Leis, J. Harbott, J. Kreuder, J. Hammermann, T. Henn and F. Lampert, Cloning and characterization of AFX, the gene that fuses to MLL in acute leukemias with a t(X;11)(q13;q23), *Oncogene* **14** (1997) 195–202.
8. T. Furuyama, T. Nakazawa, I. Nakano and N. Mori, Identification of the differential distribution patterns of mRNAs and consensus binding sequences for mouse DAF-16 homologues, *Biochem. J.* **349** (2000) 629–634.
9. F.M. Jacobs, L.P. van der Heide, P.J. Wijchers, J.P. Burbach, M.F. Hoekman and M.P. Smidt, FoxO6, a novel member of the FoxO class of transcription factors with distinct shuttling dynamics, *J. Biol. Chem.* **278** (2003) 35959–35967 (Epub 32003 Jul 35911).
10. A. Brunet, A. Bonni, M.J. Zigmond, M.Z. Lin, P. Juo, L.S. Hu, M.J. Anderson, K.C. Arden, J. Blenis and M.E. Greenberg, Akt promotes cell survival by phosphorylating and inhibiting a Forkhead transcription factor, *Cell* **96** (1999) 857–868.
11. G.J. Kops, N.D. de Ruiter, A.M. De Vries-Smits, D.R. Powell, J.L. Bos and B.M. Burgering, Direct control of the Forkhead transcription factor AFX by protein kinase B, *Nature* **398** (1999) 630–634.
12. G. Rena, S. Guo, S.C. Cichy, T.G. Unterman and P. Cohen, Phosphorylation of the transcription factor forkhead family member FKHR by protein kinase B, *J. Biol. Chem.* **274** (1999) 17179–17183.

13. E.D. Tang, G. Nunez, F.G. Barr and K.L. Guan, Negative regulation of the forkhead transcription factor FKHR by Akt, *J. Biol. Chem.* **274** (1999) 16741–16746.
14. A.M. Brownawell, G.J. Kops, I.G. Macara and B.M. Burgering, Inhibition of nuclear import by protein kinase B (Akt) regulates the subcellular distribution and activity of the forkhead transcription factor AFX, *Mol. Cell. Biol.* **21** (2001) 3534–3546.
15. D.L. Mahmud, G.M. Amlak, D.K. Deb, L.C. Platanias, S. Uddin and A. Wickrema, Phosphorylation of forkhead transcription factors by erythropoietin and stem cell factor prevents acetylation and their interaction with coactivator p300 in erythroid progenitor cells, *Oncogene* **21** (2002) 1556–1562.
16. G. Rena, Y.L. Woods, A.R. Prescott, M. Peggie, T.G. Unterman, M.R. Williams and P. Cohen, Two novel phosphorylation sites on FKHR that are critical for its nuclear exclusion, *EMBO J.* **21** (2002) 2263–2271.
17. A. Brunet, J. Park, H. Tran, L.S. Hu, B.A. Hemmings and M.E. Greenberg, Protein kinase SGK mediates survival signals by phosphorylating the forkhead transcription factor FKHRL1 (FOXO3a), *Mol. Cell. Biol.* **21** (2001) 952–965.
18. Y.L. Woods, G. Rena, N. Morrice, A. Barthel, W. Becker, S. Guo, T.G. Unterman and P. Cohen, The kinase DYRK1A phosphorylates the transcription factor FKHR at Ser329 *in vitro*, a novel *in vivo* phosphorylation site, *Biochem J.* **355** (2001) 597–607.
19. N.D. De Ruiter, B.M. Burgering and J.L. Bos, Regulation of the Forkhead transcription factor AFX by Ral-dependent phosphorylation of threonines 447 and 451, *Mol. Cell. Biol.* **21** (2001) 8225–8235.
20. M.C. Hu, D.F. Lee, W. Xia, L.S. Golfman, F. Ou-Yang, J.Y. Yang, Y. Zou, S. Bao, N. Hanada, H. Saso, R. Kobayashi and M.C. Hung, IkappaB kinase promotes tumorigenesis through inhibition of forkhead FOXO3a, *Cell* **117** (2004) 225–237.
21. M.B. Yaffe, K. Rittinger, S. Volinia, P.R. Caron, A. Aitken, H. Leffers, S.J. Gamblin, S.J. Smerdon and L.C. Cantley, The structural basis for 14-3-3: phosphopeptide binding specificity, *Cell* **91** (1997) 961–971.
22. A. Brunet, F. Kanai, J. Stehn, J. Xu, D. Sarbassova, J.V. Frangioni, S.N. Dalal, J.A. De Caprio, M.E. Greenberg and M.B. Yaffe, 14–3-3 transits to the nucleus and participates in dynamic nucleocytoplasmic transport, *J. Cell Biol.* **156** (2002) 817–828 (Epub 2002 Feb 2025).
23. C.M. Cahill, G. Tzivion, N. Nasrin, S. Ogg, J. Dore, G. Ruvkun and M. Alexander-Bridges, Phosphatidylinositol 3-kinase signaling inhibits DAF-16 DNA binding and function via 14-3-3-dependent and 14-3-3-independent pathways, *J. Biol. Chem.* **276** (2001) 13402–13410 (Epub 12000 Dec 13420).

24. W.H. Biggs, III, J. Meisenhelder, T. Hunter, W.K. Cavenee and K.C. Arden, Protein kinase B/Akt-mediated phosphorylation promotes nuclear exclusion of the winged helix transcription factor FKHR1, *Proc. Natl. Acad. Sci. USA* **96** (1999) 7421–7426.
25. D.R. Plas and C.B. Thompson, Akt activation promotes degradation of tuberin and FOXO3a via the proteasome, *J. Biol. Chem.* **278** (2003) 12361–12366 (Epub 12003 Jan 12306).
26. H. Matsuzaki, H. Daitoku, M. Hatta, K. Tanaka and A. Fukamizu, Insulin-induced phosphorylation of FKHR (Foxo1) targets to proteasomal degradation, *Proc. Natl. Acad. Sci. USA* **100** (2003) 11285–11290 (Epub 12003 Sep 11217).
27. C. Charvet, I. Alberti, F. Luciano, A. Jacquel, A. Bernard, P. Auberger and M. Deckert, Proteolytic regulation of Forkhead transcription factor FOXO3a by caspase-3-like proteases, *Oncogene* **22** (2003) 4557–4568.
28. H. Huang, D.C. Muddiman and D.J. Tindall, Androgens negatively regulate forkhead transcription factor FKHR (FOXO1) through a proteolytic mechanism in prostate cancer cells, *J. Biol. Chem.* **279** (2004) 13866–13877 (Epub 12004 Jan 13815).
29. H.H. Zhao, R.E. Herrera, E. Coronado-Heinsohn, M.C. Yang, J.H. Ludes-Meyers, K.J. Seybold-Tilson, Z. Nawaz, D. Yee, F.G. Barr, S.G. Diab, P.H. Brown, S.A. Fuqua and C.K. Osborne, Forkhead homologue in rhabdomyosarcoma functions as a bifunctional nuclear receptor-interacting protein with both coactivator and corepressor functions, *J. Biol. Chem.* **276** (2001) 27907–27912 (Epub 22001 May 27915).
30. E.R. Schuur, A.V. Loktev, M. Sharma, Z. Sun, R.A. Roth and R.J. Weigel, Ligand-dependent interaction of estrogen receptor-alpha with members of the forkhead transcription factor family, *J. Biol. Chem.* **276** (2001) 33554–33560 (Epub 32001 Jul 33502).
31. P. Li, H. Lee, S. Guo, T.G. Unterman, G. Jenster and W. Bai, AKT-independent protection of prostate cancer cells from apoptosis mediated through complex formation between the androgen receptor and FKHR, *Mol. Cell. Biol.* **23** (2003) 104–118.
32. P. Dowell, T.C. Otto, S. Adi and M.D. Lane, Convergence of peroxisome proliferator-activated receptor gamma and Foxo1 signaling pathways, *J. Biol. Chem.* **278** (2003) 45485–45491 (Epub 42003 Sep 45409).
33. K. Hirota, H. Daitoku, H. Matsuzaki, N. Araya, K. Yamagata, S. Asada, T. Sugaya and A. Fukamizu, Hepatocyte nuclear factor-4 is a novel downstream target of insulin via FKHR as a signal-regulated transcriptional inhibitor, *J. Biol. Chem.* **278** (2003) 13056–13060 (Epub 12003 Jan 13007).

34. C. Leo, H. Li and J.D. Chen, Differential mechanisms of nuclear receptor regulation by receptor-associated coactivator 3, *J. Biol. Chem.* **275** (2000) 5976–5982.
35. H.A. Tissenbaum and L. Guarente, Increased dosage of a sir-2 gene extends lifespan in *Caenorhabditis elegans*, *Nature* **410** (2001) 227–230.
36. A. Brunet, L.B. Sweeney, J.F. Sturgill, K.F. Chua, P.L. Greer, Y. Lin, H. Tran, S.E. Ross, R. Mostoslavsky, H.Y. Cohen, L.S. Hu, H.L. Cheng, M.P. Jedrychowski, S.P. Gygi, D.A. Sinclair, F.W. Alt and M.E. Greenberg, Stress-dependent regulation of FOXO transcription factors by the SIRT1 deacetylase, *Science* **303** (2004) 2011–2015 (Epub 2004 Feb 2019).
37. M.C. Motta, N. Divecha, M. Lemieux, C. Kamel, D. Chen, W. Gu, Y. Bultsma, M. McBurney and L. Guarente, Mammalian SIRT1 represses forkhead transcription factors, *Cell* **116** (2004) 551–563.
38. S.T. Henderson and T.E. Johnson, daf-16 integrates developmental and environmental inputs to mediate aging in the nematode *Caenorhabditis elegans*, *Curr. Biol.* **11** (2001) 1975–1980.
39. O. Puig, M.T. Marr, M.L. Ruhf and R. Tjian, Control of cell number by Drosophila FOXO: downstream and feedback regulation of the insulin receptor pathway, *Genes Dev.* **17** (2003) 2006–2020 (Epub 2003 Jul 2031).
40. J.M. Kramer, J.T. Davidge, J.M. Lockyer and B.E. Staveley, Expression of Drosophila FOXO regulates growth and can phenocopy starvation, *BMC Dev. Biol.* **3** (2003) 5 (Print 2003 Jul 2005).
41. M.A. Junger, F. Rintelen, H. Stocker, J.D. Wasserman, M. Vegh, T. Radimerski, M.E. Greenberg and E. Hafen, The Drosophila forkhead transcription factor FOXO mediates the reduction in cell number associated with reduced insulin signaling, *J. Biol.* **2** (2003) 20 (Epub 2003 Aug 2007).
42. T. Hosaka, W.H. Biggs, III, D. Tieu, A.D. Boyer, N.M. Varki, W.K. Cavenee and K.C. Arden, Disruption of forkhead transcription factor (FOXO) family members in mice reveals their functional diversification, *Proc. Natl. Acad. Sci. USA* **101** (2004) 2975–2980 (Epub 2004 Feb 2920).
43. N. Nakamura, S. Ramaswamy, F. Vazquez, S. Signoretti, M. Loda and W.R. Sellers, Forkhead transcription factors are critical effectors of cell death and cell cycle arrest downstream of PTEN, *Mol. Cell. Biol.* **20** (2000) 8969–8982.
44. R.H. Medema, G.J.P.L. Kops, J.L. Bos and B.M.T. Burgering, AFX-like Forkhead transcription factors mediate cell cycle regulation by Ras and PKB via p27, *Nature* **404** (2000) 782–787.
45. G.J. Kops, R.H. Medema, J. Glassford, M.A. Essers, P.F. Dijkers, P.J. Coffer, E.W. Lam and B.M. Burgering, Control of cell cycle exit and entry by protein

kinase B-regulated forkhead transcription factors, *Mol. Cell. Biol.* **22** (2002) 2025–2036.
46. M. Schmidt, S. Fernández de Mattos, A. van der Horst, R. Klompmaker, G.J. Kops, E.W. Lam, B.M. Burgering and R.H. Medema, Cell cycle inhibition by FoxO forkhead transcription factors involves downregulation of cyclin D, *Mol. Cell. Biol.* **22** (2002) 7842–7852.
47. S. Fernández de Mattos, A. Essafi, I. Soeiro, A.M. Pietersen, K.U. Birkenkamp, C.S. Edwards, A. Martino, B.H. Nelson, J.M. Francis, M.C. Jones, J.J. Brosens, P.J. Coffer and E.W. Lam, FoxO3a and BCR-ABL regulate cyclin D2 transcription through a STAT5/BCL6-dependent mechanism, *Mol. Cell. Biol.* **24** (2004) 10058–10071.
48. J. Seoane, H.V. Le, L. Shen, S.A. Anderson and J. Massague, Integration of Smad and forkhead pathways in the control of neuroepithelial and glioblastoma cell proliferation, *Cell* **117** (2004) 211–223.
49. J. Yao, E. Lai and S. Stifani, The winged-helix protein brain factor 1 interacts with groucho and hes proteins to repress transcription, *Mol. Cell. Biol.* **21** (2001) 1962–1972.
50. L. Martinez-Gac, M. Marques, Z. Garcia, M.R. Campanero and A.C. Carrera, Control of cyclin G2 mRNA expression by forkhead transcription factors: novel mechanism for cell cycle control by phosphoinositide 3-kinase and forkhead, *Mol. Cell. Biol.* **24** (2004) 2181–2189.
51. D.A. Bennin, A.S. Don, T. Brake, J.L. McKenzie, H. Rosenbaum, L. Ortiz, A.A. DePaoli-Roach and M.C. Horne, Cyclin G2 associates with protein phosphatase 2A catalytic and regulatory B′ subunits in active complexes and induces nuclear aberrations and a G1/S phase cell cycle arrest, *J. Biol. Chem.* **277** (2002) 27449–27467 (Epub 22002 Apr 27415).
52. B. Alvarez, A.C. Martinez, B.M. Burgering and A.C. Carrera, Forkhead transcription factors contribute to execution of the mitotic programme in mammals, *Nature* **413** (2001) 744–747.
53. S.R. Datta, A. Brunet and M.E. Greenberg, Cellular survival: a play in three Akts, *Genes Dev.* **13** (1999) 2905–2927.
54. S.R. Datta, H. Dudek, X. Tao, S. Masters, H. Fu, Y. Gotoh and M.E. Greenberg, Akt phosphorylation of BAD couples survival signals to the cell-intrinsic death machinery, *Cell* **91** (1997) 231–241.
55. P.F. Dijkers, R.H. Medema, C. Pals, L. Banerji, N.S. Thomas, E.W. Lam, B.M. Burgering, J.A. Raaijmakers, J.W. Lammers, L. Koenderman and P.J. Coffer, Forkhead transcription factor FKHR-L1 modulates cytokine-dependent transcriptional regulation of p27(KIP1), *Mol. Cell. Biol.* **20** (2000) 9138–9148.

56. P.F. Dijkers, K.U. Birkenkamp, E.W. Lam, N.S. Thomas, J.W. Lammers, L. Koenderman and P.J. Coffer, FKHR-L1 can act as a critical effector of cell death induced by cytokine withdrawal: protein kinase B-enhanced cell survival through maintenance of mitochondrial integrity, *J. Cell Biol.* **156** (2002) 531–542 (Epub 2002 Jan 2028).
57. P.F. Dijkers, R.H. Medemadagger, J.W. Lammers, L. Koenderman and P.J. Coffer, Expression of the pro-apoptotic Bcl-2 family member Bim is regulated by the forkhead transcription factor FKHR-L1, *Curr. Biol.* **10** (2000) 1201–1204.
58. A. Sunters, S. Fernández de Mattos, M. Stahl, J.J. Brosens, G. Zoumpoulidou, C.A. Saunders, P.J. Coffer, R.H. Medema, R.C. Coombes and E.W. Lam, FoxO3a transcriptional regulation of Bim controls apoptosis in paclitaxel-treated breast cancer cell lines, *J. Biol. Chem.* **278** (2003) 49795–49805.
59. N. Komatsu, T. Watanabe, M. Uchida, M. Mori, K. Kirito, S. Kikuchi, Q. Liu, T. Tauchi, K. Miyazawa, H. Endo, T. Nagai and K. Ozawa, A member of Forkhead transcription factor FKHRL1 is a downstream effector of STI571-induced cell cycle arrest in BCR-ABL-expressing cells, *J. Biol. Chem.* **278** (2003) 6411–6419.
60. A. Essafi, S. Fernandez de Mattos, Y.A. Hassen, I. Soeiro, G.J. Mufti, N.S. Thomas, R.H. Medema and E.W. Lam, Direct transcriptional regulation of Bim by FoxO3a mediates STI571-induced apoptosis in Bcr-Abl-expressing cells, *Oncogene* **24** (2005) 2317–2329.
61. T.T. Tang, D. Dowbenko, A. Jackson, L. Toney, D.A. Lewin, A.L. Dent and L.A. Lasky, The forkhead transcription factor AFX activates apoptosis by induction of the BCL-6 transcriptional repressor, *J. Biol. Chem.* **277** (2002) 14255–14265.
62. G.J. Kops, T.B. Dansen, P.E. Polderman, I. Saarloos, K.W. Wirtz, P.J. Coffer, T.T. Huang, J.L. Bos, R.H. Medema and B.M. Burgering, Forkhead transcription factor FOXO3a protects quiescent cells from oxidative stress, *Nature* **419** (2002) 316–321.
63. S. Nemoto and T. Finkel, Redox regulation of forkhead proteins through a p66shc-dependent signaling pathway, *Science* **295** (2002) 2450–2452 (Epub 2002 Mar 2407).
64. H. Tran, A. Brunet, J.M. Grenier, S.R. Datta, A.J. Fornace, Jr., P.S. DiStefano, L.W. Chiang and M.E. Greenberg, DNA repair pathway stimulated by the forkhead transcription factor FOXO3a through the Gadd45 protein, *Science* **296** (2002) 530–534.
65. Y. Furukawa-Hibi, K. Yoshida-Araki, T. Ohta, K. Ikeda and N. Motoyama, FOXO forkhead transcription factors induce G(2)-M checkpoint in response

to oxidative stress, *J. Biol. Chem.* **277** (2002) 26729–26732 (Epub 22002 Jun 26704).
66. J. Nakae, W.H. Biggs, III, T. Kitamura, W.K. Cavenee, C.V. Wright, K.C. Arden and D. Accili, Regulation of insulin action and pancreatic beta-cell function by mutated alleles of the gene encoding forkhead transcription factor Foxo1, *Nat. Genet.* **32** (2002) 245–253 (Epub 2002 Sep 2003).

Index